CAMBRIDGE LIBRARY COLLECTION

Books of enduring scholarly value

Physical Sciences

From ancient times, humans have tried to understand the workings of the world around them. The roots of modern physical science go back to the very earliest mechanical devices such as levers and rollers, the mixing of paints and dyes, and the importance of the heavenly bodies in early religious observance and navigation. The physical sciences as we know them today began to emerge as independent academic subjects during the early modern period, in the work of Newton and other 'natural philosophers', and numerous sub-disciplines developed during the centuries that followed. This part of the Cambridge Library Collection is devoted to landmark publications in this area which will be of interest to historians of science concerned with individual scientists, particular discoveries, and advances in scientific method, or with the establishment and development of scientific institutions around the world.

Sidereal Chromatics

Admiral William Henry Smyth's *Sidereal Chromatics* (1864) represents a landmark achievement in nineteenth-century astronomy, offering the most precise observations of the colours of double stars yet recorded. An expansion upon his well-known *Bedford Cycle of Celestial Objects*, which garnered a gold medal from the Royal Astronomical Society, *Sidereal Chromatics* provides both a theory concerning the source of double-star colours and a method for determining their most exact description. Detailed charts compare Smyth's measurements of more than one hundred double stars with his own previously published observations and those of his fellow astronomer, Father Benedetto Sestini. This edition also includes Smyth's famous colour chart, an attempt to standardise the process of identifying double-star colours. *Sidereal Chromatics* ends with Smyth's plea to amateur astronomers to continue the effort of charting the heavens, aided by improved telescopes and works such as his, 'trustworthy treatises available to all men'.

Sidereal Chromatics

Being a Re-Print, with Additions from the Bedford Cycle of Celestial Objects and Its Hartwell Continuation on the Colours of Multiple Stars

William Henry Smyth

CAMBRIDGE UNIVERSITY PRESS

Cambridge, New York, Melbourne, Madrid, Cape Town, Singapore,
São Paolo, Delhi, Dubai, Tokyo

Published in the United States of America by Cambridge University Press, New York

www.cambridge.org
Information on this title: www.cambridge.org/9781108015172

This edition first published 1864
This digitally printed version 2010

ISBN 978-1-108-01517-2 Paperback

Sidereal Chromatics.

Sidereal Chromatics;

BEING

A RE-PRINT, WITH ADDITIONS,

FROM THE

"BEDFORD CYCLE OF CELESTIAL OBJECTS,"

AND ITS

"HARTWELL CONTINUATION,"

ON THE

COLOURS OF MULTIPLE STARS.

BY

ADMIRAL W. H. SMYTH, K.S.F., D.C.L.

F.R.S., F.S.A., F.R.A.S., F.R.G.S., ETC.

LONDON:

Printed for Private Circulation by

JOHN BOWYER NICHOLS AND SONS,

PARLIAMENT STREET.

M.DCCC.LXIV.

TO JOHN LEE, ESQ., Q.C., LL.D.,

F.R.S., F.S.A., F.R.A.S., F.G.S., &c.

St. John's Lodge, 1—8—64.

My dear Friend,

Although the discoveries in sidereal astronomy since the advent of the elder Herschel, have been equally gratifying and marvellous, there is every prospect of still further triumphs for assiduity and observational skill to obtain. For this most desirable consummation, every department to which practical vigilance can render its aid, ought to be furnished to the utmost with sifted materials, in order that, from a mass of recorded facts, safe conclusions may be arrived at. This impression, together with the present number and efficiency of achromatic telescopes, induces me again to call attention to the still vague and indeterminate state in which the colours of multiple stars are registered, with a word or two as to the steps for bettering it. In our confabulations on this material point we agreed, that a reprint of my former suggestions in a portable form, would probably attract the notice and suit the convenience of many observers; the present *brochure* was therefore undertaken to carry out this view.

The method which has hitherto been pursued in these examinations is both trite and inexact, having been mostly practised when the eye, at all times a delicate organ, has been fatigued by taking and reading-off measures made with an illuminated field; and some observers register the position and distance with the magnitudes of the objects, without a word as to their colours. Now I would advise that for this purpose the objects be brought into a *dark* field, and only on very fine nights, by those *amatores scientiæ* who are possessed of powerful telescopes, but yet have no intention to labour in that most delicate branch of celestial operations, the mensuration of the angle of position of double stars and their distance from each other. In the discrimination of colours zeal and ability can render good service to the general cause, yet without the observer's encountering heavy work or toiling on monotonous reductions.

Having already stated the why and the wherefore of the question elsewhere, it only remains to show the necessity of acting in concert, and on a settled plan, so as to obtain a more exact nomenclature in *χρῶμα*. Our having noted the colours from fruits, flowers, vegetables, jewels, and the like, involves conflicting uncertainty, since most of those objects are of various tints. Indeed, as a proof of the imperfection of such comparisons, I will here cite a sample from my own practice, placing the inexact epithets which I have used in Roman print, and what was probably meant is expressed in Italics :—

Amethyst . . .	*Purple*
Apple green . . .	*Brownish green*
Ash colour . . .	*Pale dull grey*
Beet hue . . .	*Crimson*
Cinereous . . .	*Wood-ash tint*
Cherry-colour . . .	*Pale red*
Cobalt	*Bluish white*
Creamy	*Pale white*
Crocus	*Deep yellow*
Damson	*Dark purple*
Dusky	*Brownish hue*
Emerald	*Lucid green*

Fawn-coloured	*Whitey-brown*
Flushed	*Reddened*
Garnet	*Red of various shades*
Golden hue	*Bright yellow*
Grape red	*A variety of purple*
Jacinth	*Pellucid orange tint*
Lemon-coloured	*Bright but pale yellow*
Lilac	*Light purple*
Livid	*Lead colour*
Melon tint	*Greenish yellow*
Orpiment	*Bright yellow*
Pale	*Deficient in hue*
Pearl colour	*Shining white*
Plum colour	*Pale purple*
Radish tint	*Dull purple*
Rose tint	*Flushed crimson*
Ruby colour	*Pellucid red*
Ruddy	*Flesh-coloured*
Sapphire	*Blue tint*
Sardonyx	*Reddish yellow*
Sea green	*Faint cold green*
Silvery	*Mild white lustre*
Smalt	*Fine deep blue*
Topaz	*Lucid yellow*
Vanilla tint	*Dark brown or chocolate*

From various auspicious circumstances, the present appears to be a very seasonable time for bringing this interesting natural feature before our volunteer astronomers; and, as you so liberally undertook to furnish them with my hints on the subject, the recommendation before us is launched; while to it is appended a chromatic diagram, as a first step towards simplifying the records hereafter to be obtained.

This pamphlet merely touches upon the common order of colours and their complementaries—without entering into the relation of the light absorbed to the light reflected—and by no means pretends to treat of the almost infinite shades and tints which philosophers and artists distinguish in nature. The recent discovery of a mono-chromatic green light occasioned by the combustion of the new and curious metal Thallium, establishes the Newtonian theory of the spectrum, as being more correct than the ostensible dogma of only three primary colours, which had nearly supplanted it of late; while the mysterious lines within the arch of seven primitively refracted rays, open out a wonderful scope in the development of planetary and sidereal physics. Indeed the rigorous spectrum analysis now in hand, as applied to the stars—albeit of operose difficulty—is rich in promise; and already do the observations of Kirchoff, Donati, and Rutherford, with the masterly experiments of Miller and Huggins, indicate that a cause for the difference of colour may thus be sought in the various constitutions of the investing stellar atmospheres.

But, even were those beautiful inquiries more matured and correlative than they are as yet, the call now made to the efficiently-armed amateur would not be the less necessary, seeing what a series of well observed and accurately described facts may be reaped by diligent appliance.

You must well remember the attempt that we made in July 1829, hereafter recounted, to come to a conclusion regarding the colours of the double-star familiarly designated Cor Caroli; and how eleven pairs of eyes, of which six couple were directed by ladies, widely differed in their mental appreciation of the hues before them. Yet in the autumn of 1855, myself and two other observers were unanimous in pronouncing that the same object was—

A. Pale reddish white. B. Lilac;

the which might have been more satisfactorily and

more exactly expressed according to the diagram herein proposed, in this form :—

A. Red^{4}. B. Purple^{3}.

At this last epoch, by direct experiment on the mild tints of Cor Caroli and the stronger ones of Rasalgeti, I was satisfied that, however telescopes may vary as to the intensity of the hue imparted, according to the accuracy of their chromatic correction, and even under different powers by the same instrument, yet the general tone is in fair accordance. It follows then, that the adoption of the proposed line of action, till a better one is presented, will assuredly be productive of many desirable results; and therefore, hoping for a realization of the wish, I am always

Yours very truly,

CONTENTS.

PART I.

ON THE COLOURS OF DOUBLE STARS.

UNDER this title we are not about to inflict on the reader any of the extravagant notions of the ancients—aye, not even the best of them—concerning the essence of hues and tints, they being utterly unsupported either by reason or experiment. Nor is it intended to theorize under modern advantages upon light and rays, the seductions of harmony and contrast, nor to jump at any of the remaining uncertainties of the question. The present light sketch is merely handed forth to investigate that property of colours which affects the sight only: and, if the absolute tints of double stars are determined at special epochs, much will be contributed towards the mysterious laws which regulate such variations as may be observed. Upon this understanding, we will open our fire with the following extract from the Prolegomena to my

The Author's intent.

CYCLE OF CELESTIAL OBJECTS.

For the sake of the tyro, it may be proper to premise that the light of the two stars composing the double one is frequently found to be of very dissimilar intensity, and often of very dissimilar colours; and no one who has ever directed a telescope to the heavens can have failed to be struck with the brilliant hues they present, especially such lovely objects as γ Andromedæ, α Herculis, and ε Boötis. The phenomenon of the tints displayed by the smaller stars is considered by M. Arago as owing to an excess of refrangible rays, acted upon by an absorbing force in the atmosphere of the larger star; but this cannot be the universal law. Sometimes the strongest of the two is of a yellow, red, or orange tinge, still more frequently

Initiatory Remarks.

the secondary is blue, purple, or greenish, and those colours so palpable as to be visible in objects of the smallest magnitude. Now, as many of these pairs border on the extremes of the prismatic spectrum, the larger star being allied to the red, and the smaller to the violet, the exhibition may in such cases be the effect of contrasted complementary tints—corresponding to the male and female lights of Milton.* We all know that a white light appears greenish when near a strong red one, and becomes blueish when the neighbouring colour is yellow. In combinations of this nature some of the secondaries lose their colour on hiding the primary; but, as many of the examples defy this test, their colours are too decidedly indicated to be merely imaginary. For instance, as *a* Leonis is of a brilliant white tint, the deep purple of its *comes* cannot be an illusion, and in δ Serpentis both the bodies are blue.

Struve's Estimates.

It will thus be seen that this department of sidereal inquiry offers an interesting field for continuous investigation, and accordingly we find that the elder Struve has paid strict attention to it. By his observations it is confirmed that, besides the white, there exist stars of every shade of the prism; and that, when the principal body is not white, it approaches the red side of the spectrum, while the satellite offers the bluish of the opposite extreme. Yet this apparent law is not without exceptions in the Catalogue before us; on the contrary, the most general case is that the two stars are of the same colour, as will be seen in the following summary, wherein he finds among 596 brilliant double stars:

375 pairs of the same colour and intensity.
101 pairs of the same colour, but different intensity.
120 pairs of totally different colours.

---

* On this head Sir John Herschel has remarked, that it is "probably in virtue of that general law of optics which provides that when the retina is under the influence of excitement by any bright-coloured light, feebler lights, which when seen alone would produce no sensation but of whiteness, shall for the time appear coloured with the tint complementary to that of the brighter."

Among those of the same colour, the white are the most numerous, and of 476 specimens of that species he found:

295 pairs, both white.
118 pairs both yellowish, or reddish.
63 pairs, both bluish.

Therefore the number of red or reddish stars is double that of the bluish tinge, and that of the white stars is twice and a half greater than the red ones. The combination of a blue companion with a coloured primary happens:

53 times, with a white principal star.
52 times, with a light yellow.
52 times, with a yellow or red.
16 times, with a green.

Obstacles to concurrence.

Professor Struve's chromatic designations are obscurissima, obscura, pallida, livida, alba, sub-flava, flava, sub-cærulea, cærulea, rubicunda, and rubra.* He supposes the ninth magnitude to be the outside boundary in which he recognises colour, but I have been much struck with the beautiful blue tint of several of the smallest stars visible in my telescope. This, however, may be attributed to some optical peculiarity. The Professor found what I have also experienced, that Sir William Herschel saw most objects with a redder tinge than they have since proved to be of. This may be owing to the effect of his metallic mirror or to some peculiarity of vision, or perhaps both. We know there are many examples of very sharp eyes being unable to distinguish colours correctly, among whom may be instanced the late George Duke of Marlborough, who was an amateur astronomer, and possessed a good sight in other respects. There are others who have this singular physical defect with regard to particular colours only, as our estimable countryman Dalton, who, though so conversant with the laws of the spectrum, could not discriminate between scarlet and

* That masterly observer, perplexed by the tint of the close companion of ξ Orionis, somewhat sesquipedalianly designates it *olivaceasubrubicunda*. I entered it as light purple, and Sestini called it azure.

brown. So also those two celebrated men, Troughton and Dugald Stewart, were affected, but their peculiarity of vision consisted in confusing scarlet with green, and pink with blue. To the former the ripe cherry and its leaf were of one hue, only to be distinguished by their form, yet his eyesight was sharp enough for the examination of the minutest sub-divisions upon graduated instruments. Among other instances of this peculiarity of the sense of colours, I was greatly surprised on finding that an intimate friend of my own (Sir G. Musgrave) could not perceive the strong colours of γ Andromedæ and other remarkable stars with my telescope, as I was well aware of his exquisite taste and execution in missal-blazonry; but he assured me that he cannot easily discriminate between brown and green; albeit I have specimens of his art in which those colours are treated as well as they could be by any one in the Royal Academy. Sir John Herschel examined the eyes of an individual thus circumstanced, and satisfied himself that all the prismatic rays had the power of affecting them with the sensation of light, and producing distinct vision; so that he considers this defect as arising from no insensibility of the retina to rays of any peculiar refrangibility, but rather as residing in the sensorium, by which it is rendered incapable of appreciating exactly those differences between the rays on which these colours depend.*

Difficulty of deduction.

The permanency or variation of star-tints requires still closer observation before any correct deductions can be drawn. The ancients recognised no blue stars; they only spoke of white or red ones, classing among the latter Arcturus, Aldebaran, Pollux, Antares, and Betelgeuze, which also appear red to us; but they added Sirius, the *rubra canicula* of the poets, which, though expressly declared to be red both by Ptolemy and Seneca, is now decidedly white, and brilliantly white too. This instance affords a strong presumption that these colours undergo

---

* An intelligent seaman, Captain Joseph Huddart, wrote "Of persons who could not distinguish colours."—See the *Philosophical Transactions* for 1777.

changes; there are two remarkable examples of a recent date in γ Leonis and γ Delphini, which, at Sir William Herschel's time of observing them, seem to have been perfectly white, though the first pair is now of a golden yellow and reddish green, and the second of a bright yellow and bluish green. Blue stars are of modern introduction, since they are first mentioned by Mariotte in 1686, who supposes that they owe this colour to their being fainter, and free from exhalations. Mr. Dunlop, in the Catalogue made at Paramatta, in Sir Thomas Brisbane's observatory, mentions a large group of stars, all the individuals of which are blue—also a bluish nebulosity; but we have no such object in the northern hemisphere. Nor is it less remarkable that amidst this infinite variety of tints, although single red stars are frequently met with, there is not an instance of a solitary green, purple, blue, or violet-coloured one being found; and, among other singularities, the irregularity and even absence of the principal prismatic shades during the phases of most of those stars that vary in their magnitude, is a striking phenomenon, whence may be yet deduced important conclusions respecting the velocity of different coloured rays.

Sir J. Herschel on coloured suns.

Herschel II. treating of contrasted colours in his admirable *Outlines of Astronomy*, finely remarks:—"It is by no means, however, intended to say, that in all such cases one of the colours is a mere effect of contrast, and it may be easier suggested in words than conceived in imagination, what variety of illumination *two suns*—a red and a green, or a yellow and a blue one—must afford a planet circulating about either; and what charming contrasts and 'grateful vicissitudes'—a red and a green day, for instance, alternating with a white one and with darkness—might arise from the presence or absence of one or other, or both, above the horizon." Indeed the beautiful effects of such glorious prismatic effulgence among the probable wanderers, offers a tempting arena to the fanciful contemplation of imaginative powers.

Notions of old

Zahn—a zealous opponent of the Copernican system—in his *Syntagma*, remarks that the stars shine " more like torches burning with eternal flame before the altar of the Most High, than the lamps of the ætherial vault, or the funeral lights of the setting sun;" and he descants on their colours, asserting that from the various hues of the fixed stars, their nature may be inferred, and the planets they imitate at once known. Hence some are designated *Saturnine*, some *Jovial*, and others *Martial*. The Saturnine are those of a leaden or livid colour, and dullish; the Jovial are bright and white; and the rusty-coloured ones are assigned to Mars. The Solar ones, partly yellow and partly red, shine very splendidly; those of Venus are of a box-coloured glow—*Venereæ sunt buxeæ, seu buxeo splendore clarescunt;* and the Lunar are pale and dim. Such was the knowledge of stellar colours in 1694; a century and a half has done somewhat to alter all this.

Arago on Green and Blue Stars.

Our redoubtable colleague, Arago, treating of this interesting subject in his *Astronomie Populaire,* says: " The phenomenon of the colour of multiple stars has been observed for too short a time to expect a plausible explanation of it in the present day. It will be reserved for time and precise observations to inform us whether green or blue stars are not suns already in course of decay. If the different shades of those stars do not indicate a process of combustion in different stages: if the tinge with an excess of the more refrangible rays which the smaller star frequently exhibits, does not arise from the absorbing power of an atmosphere which might be developed by the action of the usually more brilliant star which it accompanies," &c. Euge! Now here is blue ruin with such a vengeance that we can but marvel at the temerity of a worthy *savant* in ventilating so hasty and wild a supposition; but theory, especially when unsupported by crucial or instantial facts, is mighty prone to treading upon hot ashes.

Variable Stars.

However, in our present thirst for exact knowledge, it will

be as well to admit the reasonableness of the De Coverley maxim, in that there is something to be said on all sides, for, since the foregoing remarks were first printed, much has been achieved among the Variable Stars, as well those which are proved to be periodical in their changes of light or brilliancy, as others which are yet only to be termed versatile. It is a promising field for exploration, even though the physical nature of the fluctuating bodies may continue to remain a latent mystery; at present they must be classed with those postulates regarding which Sir John Herschel pithily observes, " it is easier to propound questions than to offer satisfactory replies to them."

Hind's ruddy tinted stars.

My indefatigable friend Mr. Hind, who has worked to admiration among the Variables, has noted that these stars, especially the fainter ones, have generally a ruddy tint; upon which Mons. Arago thus dilates:—" Might there not be some connexion between this remark and the observation made by the same astronomer, that variable stars at the instant of their minimum brightness appear surrounded with a kind of fog? Supposing the existence of this fog well established, we should be on the road to the explanation of these singular phenomena. Perhaps we may arrive at the conclusion that a star's variations of brightness are due, not to a perfectly opaque planet revolving round it, but to cosmical clouds, which, by a similar movement of revolution, would be successively interposed between those bodies and the earth."

Now whatever may hereafter prove to be the upshot of our " opinions quaint," it must be received as a crucial fact in the discussion, that nine of Hind's Variables are recorded as being red, and one absolutely *crimson,* like to " a blood-drop on the black ground of the sky." The fog or haze at the minimum was palpably obvious, and in several some very decided changes of colour took place at different stages of their light-curve, for instance, R Geminorum, announced by its discoverer as passing through blue, yellow, and red, during the 371 days of the gradations in its lustre, which is fully confirmed by the accurate Mr. Norman Pogson. Besides the detection of the above-mentioned crimson star, Mr. Hind thus wrote me from Regent's Park under date of the 14th January, 1850:—

I also avail myself of this opportunity to mention that, in October 1845, I remarked a most fiery or scarlet star on the confines of Lepus and Orion, Æ (1850) 4 h. 52 m. 45 s. and N P D 105° 2′. This is *by far* the most deeply-coloured of any that I have yet seen, and in striking contrast with a beautifully white star preceding it one minute. It is not inserted in Sir John Herschel's Catalogue of Red Stars, there is no allusion to it in your "Cycle," nor can I find any previous notice of it; yet I am doubtful whether I can have first discovered this remarkable star.

Hind's Mira Ophiuchi.

The same diligent observer on the night of the 28th of April, 1848, suddenly perceived a red star of between the fourth and fifth magnitude, very conspicuous to the naked eye, in the region of the Serpent-bearer, in Æ 16 h. 51 m., and in Dec. — 12° 39′, where, from a thorough acquaintance with the spot, he felt assured, and could state positively, that no star down to the tenth magnitude had previously existed. It was to unassisted vision as bright as ν Serpentis. I happened to be absent from home when Mr. Hind sent me notice of this "Mira Ophiuchi," nor was it till the end of June that I was able to get a telescopic glimpse of it when it had dwindled to about the eighth degree of lustre only. From the time of its discovery the stranger continued to diminish, but without altering its position; and, before the advance of the season rendered observations impracticable, it had nearly disappeared. But, although it has been watched from 4·5 to below 13·5 magnitude, it cannot as yet be enrolled among the periodical Variables, and for the present must remain under the designation of *Hind's Changing Star.*

In the present stage of inquiry—an inquiry of no small moment in universal physics—the Variables can hardly be placed in the same category with the Double Stars to which I am drawing attention; for, though some of them may be cosmically changeable in colour, the conditions must be essentially different between a single star like the above-cited R Geminorum, ranging to our eyes from the seventh to the eleventh degrees of brightness, and such a steady bold double object as 95 Herculis—for which see the Appendix. Both classes, however, are proved to be entities, and it remains to watch the Variables through their curves of light with unremitting diligence, in order to arrive at more satisfactory results. Some of the temporary phenomena, noted in low stars, may be caused by the mutable conditions of the earth's atmosphere and the interference of light; but it is certain that most of the variations above alluded to, are entirely independent of any effect which the said atmosphere could produce.

Variable Star Atlas.

In the *Speculum Hartwellianum*, page 128, it is shown that on engaging Mr. N. R. Pogson as Dr. Lee's astronomical assistant, an express condition was, that a special scrutiny be made into the desiderata of those wondrous bodies relative to which we have just spoken; and such was his diligence that, within the short space of two years, twenty maps of known or suspected Variables, together with the chief recorded Temporary stars, were in so advanced a state of preparation as to be nearly ready for engraving. But, being suddenly called away, in order to assume the Directorship of the important Observatory at Madras, he carried the results of his former labours with him, to superintend their due completion; when, with the express permission of Government, the work will be published under the title of HARTWELL AND MADRAS ATLAS OF VARIABLE STARS.

PART II.

DOUBLE STAR COLOURS CONTINUED.

Opening remarks.

IN consecution of this fascinating subject, we will now follow the period of 1844, the date of the "Cycle," by that of 1860, the year in which the Hartwell Continuation appeared; after a further advertence to most of its matter, and considerable correspondence on the several points connected therewith. It has been suggested to me that some colours may undergo pulsations, but the adduced instances are mostly at low altitudes, where atmospheric influences are prevalent; and there may be a want of rigorous correction of the residual spectrum of the refracting telescope. But embarrassments in the outset of any enterprize only enhance the value of proven examples.

It may be noted—*en passant*—that a slight perturbation pervaded the minds of observing neophytes on the averment pronouncing that there are only three primary colours, namely—red, yellow, and blue; and that the other four—orange, green, indigo, and violet—are *de facto* produced by combinations of the former, and are therefore secondary or compound colours. This is, however, comparatively easy, as well to suggest as to adopt; but when an arrogant Goethe—unversed even in first principles—steps forward in the pride and panoply of popularity to explain the physiological and chemical qualities of the same, in order to demolish the "nauseous precepts of Newton," we are really taken aback by his temerarious effrontery. Only think, that because the artists in mosaic at Rome are said to employ 15,000 varieties of hues—each of which has 50 shades from the darkest to the lightest—he would saddle us with 750,000 different tints; and be it observed, the workmen in mosaic use opaque colours only. The intellectual giant had no business to go beyond his proper literary bounds, and poach upon science.

In now resuming our theme, we will submit a reprint, with increment, of the VIIth Chapter of the—

SPECULUM HARTWELLIANUM.

First the flaming red
Sprung vivid forth; the tawny orange next;
And next delicious yellow; by whose side
Fell the kind beams of all-refreshing green.
Then the pure blue that swells autumnal skies,
Ethereal play'd; and then of sadder hue,
Emerg'd the deepen'd indigo, as when
The heavy-skirted evening droops with frost.
Whilst the last gleamings of refracted light
Dy'd in the fainting violet away.

THOMSON.

The Sestini Comparisons.

In the Cycle of Celestial Objects (vol. i. page 300-4) I have dwelt generally upon the colours of double stars; and in the Bedford Catalogue, which forms the second volume of that work, have assigned such colours to all the objects as struck me at the time of observation. It seems that towards the close of the labours of Signor Benedict Sestini, of Rome, on a very extensive catalogue of stars, my book reached his hands, and led him to form the table which he has appended to his volume,* in hopes that the colours of stars may be more strictly watched. The conclusions which he had arrived at were, that of two thousand five hundred and forty stars (those of Baily's Catalogue observed at Rome) the yellow stars are about half the total number, and equally distributed; the white stars are one-fifth, in scattered portions; and the orange rather more than one-fifth. The red and the blue are rare from the Pole to 30° of north declination; the blue then become numerous ($= \frac{1}{7}$) to the Equator, especially from

* A selection from these, twenty nine in number, was communicated by the late Professor de Vico to Schumacher at Altona, in 1848; who published the details in No. 684 of the Astronomische Nachrichten.

Æ 18^h to 20^h; and the red abound from 0° to 30° south declination, and Æ 16^h to 20^h.

Sestini's method.

On Signor Sestini's instituting the comparisons which he has described in his work, he carefully recorded the dates, as a step to ascertaining whether the colours may or may not be found to vary with time. Political furor—a deadly foe to science—drove De Vico, Sestini, and other distinguished members of the Collegio Romano, from their country; and the two former sought and found an asylum in America. In passing through London, Professor De Vico had given me notice of the use to which my Cycle had been applied; and I afterwards received a copy of Sestini's work, with the following letter from the worthy author in English, dated George-town College, March 19th, 1849:

Soon after I had completed this collection of observations, I conceived the design of sending a copy to you, not only as a mark of respectful esteem, but also from having profited by your Celestial Cycle in the arrangement and comparison of the observations that I myself had made in the Observatory of the Roman College. But the unmerited expulsion and exile that I and my brethren have suffered, have obliged me to leave almost everything, and the printed copies of the little *Memoria* remained in the printer's hands, where I think they are yet. Luckily, a while ago I found a few copies in a trunk, sent to me after the death of our dear De Vico. Therefore I send you now what I could not send you before, and at the same time, and without troubling you, I would desire to know what you advise upon this sort of observations. And, as I have the same objective glass of Cauchoix that I have used in Rome to collect these observations, I desire to know if you approve a review, not entirely but sufficiently general in order to examine if the difference of the climate have any influence on the colour of stars. You can also, if you please, suggest more advisable means than those I have adopted, and I should be very happy if the opportunity of having the same and so good an objective glass that I have used for ten years in Italy could give me the means of lending some advantage to the science: what so sincere a lover of this beautiful science as you are can easily find. On another occasion, and when I become better acquainted with the English language, I hope to thank you for all your favours and kindness. Then also will I tell something of the new observatory of this George-town College, erected and very well furnished by the care of M. Curley, a very able and industrious professor, my colleague.

The observations made and published with my first *Memoria* are reprinted and collected together with all the others of this second *Memoria*

Nevertheless, if I had here a copy, I would join to the second, and I would send both to you.

Remarks to Sestini.

In reply to Signor Sestini, I expressed my satisfaction with the course of his inquiries, as they might yet be of great value in a very interesting branch of physical phenomena. He should, however, remember that the colours recorded in the Cycle were frequently noted after the eye was fatigued and biassed by previous working in an illuminated field, and reading minute divisions on graduated micrometer-circles. There were also the imperfections of the sight, the materials of the object-glass, and the various atmospheric media to be considered before any crucial exactness could be expected; but that still, under every objection, approaches to satisfactory conclusions must inevitably follow stricter attention to the subject. I therefore recommended occasional references to the heavens for this object only, with the eye kept in sharp order for the purpose, having carefully tested the capacity of that organ in strictly identifying the several gradations of colour. Many of the tints of stellar companions would of course turn out to be merely complementary contrasts; but the inherent colours would become additionally valuable, as strict observation upon them, under efficient means, advanced. Indeed, I added, it may be considered as fully proved, that the hue of some stars is not the effect of contrast, but a physical reality.*

Death of De Vico.

The unfortunate exiles—De Vico and Sestini—obtained a refuge in George-town, on the banks of the Potomac, in the United States; hoping for employment in the newly-erected observatory there. Shortly afterwards, in furtherance of instruments and other scientific objects, the former recrossed the Atlantic Ocean; but, worn with fatigue and anxiety of mind, he died in London on the 15th of November, 1848.

* Secchi (*Osservatorio del Collegio Romano*) says —"Until now a means has been wanting to decide the degree of colour of double stars with precision. If I do not deceive myself I have succeeded, at least to indicate a means by which we can determine it without mistake in centuries to come: this means is derived from the spectrum of the electric spark (*questo mezzo è desunto dallo spettro della scintilla elettrica*), and I will shew it in conclusion. My attempts hitherto, for want of time and apparatus, have not been reduced to practice."

Though usually known as a successful comet-hunter, De Vico was alive to all other branches of celestial research; and he had promised me that, whenever opportunity offered, he would follow up "i colori insigni delle stelle."

Ladies' aid called in.

The question is pregnant with interest; and, considering that it may be advantageously encountered by any diligent, sharp-sighted amateur who, possessed of a good telescope and inclined only to easy and pleasing work, is nevertheless zealous to become useful in the cause of knowledge, I have herewith subjoined the list of chromatic comparisons of Sestini's observations with mine; to which are added those colours which I have since noted at Hartwell. Several of them were confirmed in direct comparison by Dr. Lee; it was an occasion on which the eyes of ladies also were prized, and the tints were mostly pronounced without reference to the older records. Here follow the details:—

Stars.	Cycle No.	SMYTH.		SESTINI.		SMYTH.	
		Epoch.	Colours.	Epoch.	Colours.	Epoch.	Colours.
35 Piscium . . .	7	1837·9	*A. Pale white* *B. Violet*	1844·8	*A. Yellowish* *B. Azure*	1850·7	*A. White* *B. Purplish*
113 P. O. Ceti . . .	17	1832·8	*A. Cream yellow* *B. Smalt blue*	1845·8	*A. Yellowish* *B. Azure*	1849·7	*A. Yellowish* *B. Fine blue*
146 P. O. Ceti . . .	23	1837·9	*A. Pale topaz* *B. Violet*	1845·8	*A. Orange yellow* *B. Dull azure*	1849·7	*A. Yellow* *B. Flushed blue*
η Cassiopeæ . . .	29	1843·2	*A. Pale white* *B. Purple*	1845·6	*A. Yellow* *B. Orange*	1850·6	*A. Dull white* *B. Lilac*
65 Piscium	31	1838·2	*A. Pale yellow* *B. Pale yellow*	1844·8	*A. Yellowish* *B. Azure*	1850·8	*A. Pale yellow* *B. Pale yellow*
ψ Piscium	37	1833·9	*A. Silvery white* *B. Silvery white*	1844·8	*A. Fine azure* *B. Fine azure*	1849·7	*A. Flushed white* *B. Pale white*
α Ursæ Minoris . .	44	1838·2	*A. Topaz yellow* *B. Pale white*	1845·6	*A. Yellowish* *B. Azure*	1849·6	*A. Yellow* *B. Dull white*
ζ Piscium	47	1839·0	*A. Silver white* *B. Pale grey*	1844·8	*A. Yellow* *B. Dingy yellow*	1849·7	*A. White* *B. Greyish*
37 Ceti	48	1838·9	*A. White* *B. Light blue*		*A. Yellow* *B. White*	1849·7	*A. Creamy white* *B. Dusky*
ψ Cassiopeæ . . .	52	1836·3	*A. Orange tint* *B. Blue*		*A. White* *B. White*	1850·2	*A. Golden yellow* *B. Ash-coloured*
85 P. I. Piscium . .	54	1837·0	*A. Yellow* *B. Pale blue*	1844·8	*A. Yellow* *B. Azure*	1855·4	*A. Pale yellow* *B. Bluish*
γ Arietis	72	1837·9	*A. Bright white* *B. Pale grey*	1844·9	*A. White* *B. White*	1850·7	*A. Full white* *B. Faint blue*
λ Arietis . . .	76	1830·9	*A. Yellowish white* *B. Blue*	1844·9	*A. White* *B. Pale azure*	1857·7	*A. Pale yellow* *B. Flushed blue*

Stars.	Cycle No.	SMYTH.		SESTINI.		SMYTH.	
		Epoch.	Colours.	Epoch.	Colours.	Epoch.	Colours.
α Piscium	**81**	1838·9	*A. Pale green* *B. Blue*	1844·8	*A. White* *B. White*	1850·8	*A. Greenish* *B. Pale blue*
γ Andromedæ . .	**82**	1843·3	*A. Orange* *B. Emerald green*	1846·5	*A. Red orange* *B. Lighter red*	1850·3	*A. Deep yellow* *B. Sea-green*
14 Arietis	**86**	1833·9	*A. White* *B. Blue*	1844·9	*A. Yellowish* *B. Bluish white*	1857·7	*A. Pale white* *B. Grey*
72 P. II. Cassiopeæ .	**97**	1834·8	*A. Pale yellow* *B. Lilac*	1845·6	*A. White* *B. White*	1857·7	*A. Yellowish white* *B. Purplish*
θ Persei	**109**	1833·6	*A. Yellow* *B. Violet*	1845·6	*A. Yellow white* *B. Azure*	1849·6	*A. Yellow* *B. Dusky blue*
π Persei . . .	**115**	1838·8	*A. Orange* *B. Smalt blue*	1845·7	*A. Golden orange* *B. Azure*	1850·7	*A. Reddish yellow* *B. Blue*
32 Eridani	**147**	1843·2	*A. Topaz yellow* *B. Sea-green*	1845·9	*A. Yellow* *B. White*	1850·3	*A. Bright yellow* *B. Flushed blue*
φ Tauri	**158**	1832·8	*A. Light red* *B. Cerulean blue*	1845·8	*A. Golden orange* *B. Azure*	1852·5	*A. Pale red* *B. Blue*
χ Tauri	**160**	1831·9	*A. White* *B. Pale sky-blue*	1845·8	*A. White* *B. Azure*	1850·7	*A. White* *B. Grey*
62 Tauri	**161**	1835·9	*A. Silver white* *B. Purple*	1845·8	*A. White* *B. White*	1850·7	*A. White* *B. Pale purple*
88 Tauri	**169**	1832·9	*A. Bluish white* *B. Cerulean blue*	1845·8	*A. White* *B. White*	1852·5	*A. Bluish white* *B. Blue*
τ Tauri	**171**	1831·9	*A. Bluish white* *B. Lilac*	1845·8	*A. Very white* *B. Azure*	1852·5	*A. Pale white* *B. Violet*
ω Aurigæ	**174**	1833·8	*A. Pale red* *B. Light blue*	1845·7	*A. White* *B. White*	1850·7	*A. Flushed white* *B. Light blue*
62 Eridani	**175**	1831·9	*A. White* *B. Lilac*	1845·9	*A. Light yellow* *B. Azure*	1852·5	*A. Pale white* *B. Flushed blue*
14 Aurigæ	**188**	1832·8	*A. Pale yellow* *B. Orange*	1845·7	*A. White* *B. Azure*	1850·7	*A. Greenish yellow* *B. Bluish yellow*
23 Orionis	**197**	1835·2	*A. White* *B. Pale grey*	1845·9	*A. Yellowish* *B. Bluish white*	1850·2	*A. Creamy white* *B. Light blue*
111 Tauri	**198**	1832·9	*A. White* *B. Lilac*	1845·9	*A. Yellowish* *B. White*	1857·7	*A. Pale white* *B. Lilac*
118 Tauri	**205**	1838·9	*A. White* *B. Pale blue*	1845·9	*A. White* *B. White*	1850·2	*A. White* *B. Bluish*
δ Orionis	**211**	1835·1	*A. Brilliant white* *B. Pale violet*	1845·9	*A. Yellowish* *B. Very white*	1850·2	*A. Pale white* *B. Flushed white*
λ Orionis	**215**	1843·1	*A. Pale white* *B. Violet*	1845·9	*A. Yellowish* *B. Bluish white*	1850·2	*A. Pale yellow* *B. Purplish*
ι Orionis	**218**	1832·1	*A. White* *B. Pale blue*	1845·9	*A. Slightly yellow* *B. Azure*	1852·5	*A. Pale white* *B. Bluish*
26 Aurigæ	**220**	1833·1	*A. Pale white* *B. Violet*	1845·7	*A. Yellowish* *B. Blue*	1849·7	*A. Dusky white* *B. Pale blue*
σ Orionis	**222**	1832·2	*A. Bright white* *B. Bluish*	1845·9	*A. Yellow* *B. Azure*	1850·3	*A. White* *B. Grey*
ζ Orionis	**223**	1839·2	*A. Topaz yellow* *B. Light purple*	1845·9	*A. Yellowish* *B. Azure*	1850·3	*A. Yellow* *B. Flushed blue*
γ Leporis	**225**	1832·1	*A. Light yellow* *B. Pale green*	1845·9	*A. Orange yellow* *B. Orange red*	1852·2	*A. Pale yellow* *B. Flushed*

Stars.	Cycle No.	SMYTH.		SESTINI.		SMYTH.	
		Epoch.	Colours.	Epoch.	Colours.	Epoch.	Colours.
8 Monocerotis . .	**245**	1834·2	*A. Golden yellow* *B. Lilac*	1845·9	*A. Pale yellow* *B. Yellowish*	1850·8	*A. Yellow* *B. Flushed blue*
15 Geminorum . .	**247**	1832·0	*A. Flushed white* *B. Bluish*	1845·9	*A. Orange* *B. Yellowish*	1852·4	*A. Pale white* *B. Ash-coloured*
20 Geminorum . .	**252**	1834·0	*A. Topaz yellow* *B. Ceruleun blue*	1845·9	*A. Yellowish orange* *B. Yellow*	1849·7	*A. Yellow* *B. Pale blue*
π^2 Canis Majoris . .	**270**	1834·1	*A. Flushed white* *B. Ruddy*	1845·9	*A. Yellowish* *B. Reddish*	1851·3	*A. Bluish white* *B. Ruddy*
α Geminorum . .	**292**	1843·1	*A. Bright white* *B. Pale white*	1845·9	*A. Yellowish* *B. Yellow*	1849·2	*A. Very white* *B. Pale white*
ζ Cancri	**315**	1843·1	*A. Yellow* *B. Orange tinge*	1846 0	*A. Yellow* *B. White*	1849·2	*A. Yellow* *B. Bright yellow*
ϕ^2 Cancri	**320**	1843 2	*A. Silvery white* *B. Silvery white*	1846·0	*A. Yellowish* *B. White*	1849·2	*A. White* *B. Pale white*
υ' Cancri	**321**	1843·2	*A Pale white* *B. Greyish*	1846·0	*A. White* *B. White*	1849·2	*A. White* *B. Dusky white*
72 P. VIII. Argo Navis	**323**	1830·8	*A. Red* *B. Green*	1846·1	*A. Orange red* *B. Yellow*	1851·3	*A. Orange* *B. Bluish green*
108 P. VIII. Hydræ .	**326**	1839·1	*A. Pale yellow* *B. Rose tint*	1846·0	*A. Orange* *B. Orange*	1849·2	*A. Full yellow* *B. Flushed*
ι Cancri	**336**	1836·2	*A. Pale orange* *B. Clear blue*	1846·0	*A. Fine orange* *B. Azure*	1851·3	*A. Dusky orange* *B. Sapphire blue*
τ' Hydræ	**360**	1831·9	*A. Flushed white* *B. Lilac*	1846·1	*A. Yellow* *B. Yellow*	1851·3	*A. Pale white* *B. Dusky*
6 Leonis	**363**	1832·2	*A. Pale rose tint* *B. Purple*	1846·0	*A. Fine orange* *B. White*	1851·3	*A. Flushed yellow* *B. Pale purple*
7 Leonis	**364**	1832·2	*A. Flushed white* *B. Violet tint*	1846·0	*A. Rather yellow* *B. White*	1851·3	*A. Bluish white* *B. Pale violet*
9 Sextantis . . .	**371**	1832·2	*A. Blue* *B. Blue*	1846·0	*A. Dingy orange* *B. Dingy orange*	1851·3	*A. Flushed blue* *B. Pale blue*
35 Sextantis . . .	**384**	1839·1	*A. Topaz yellow* *B. Smalt blue*	1846·1	*A. Pale yellow* *B. Pale yellow*	1849·2	*A. Rich yellow* *B. Ceruleun blue*
54 Leonis	**391**	1839·3	*A. White* *B. Grey*	1846·0	*A. Yellow* *B. White*	1851·3	*A. Silvery white* *B. Ash-coloured*
ϕ Leonis	**405**	1831·2	*A. Pale yellow* *B. Violet*	1846·2	*A. Pale yellow* *B. White*	1851·3	*A. Pale yellow* *B. Dusky red*
90 Leonis	**421**	1835·4	*A. Silvery white* *B. Purplish*	1846·1	*A. White* *B. White*	1851·3	*A. Silver white* *B. Pale purple*
δ Corvi	**446**	1831·3	*A. Pale yellow* *B Purple*	1846·3	*A. Slightly yellow* *B. White*	1851·3	*A. Light yellow* *B. Purple*
24 Comæ Berenicis .	**451**	1836·4	*A. Orange colour* *B. Emerald tint*	1844·4	*A. Gold* *B. Azure*	1851·3	*A. Orange* *B Lilac*
143 P. XII. Virginis .	**453**	1833·3	*A. Pale yellow* *B. Greenish*	1846·3	*A. Red* *B. Azure*	1851·3	*A. Yellowish* *B. Flushed blue*
12 Canum Venaticorum	**466**	1837·4	*A. Flushed white* *B. Pale lilac*	1844·5	*A. Yellow* *B. Blue*	1850·5	*A. Full white* *B. Very pale*
ζ Ursæ Majoris . .	**480**	1839·3	*A. Brilliant white* *B. Pale emerald*	1844·5	*A. White* *B. Yellowish*	1849·2	*A. White* *B. Pale green*
ι Bootis	**508**	1838·2	*A. Pale yellow* *C. Creamy*	1844·5	*A. Orange yellow* *C. Azure*	1850·6	*A. Light yellow* *C. Dusky white*

Stars.	Cycle No.	SMYTH.		SESTINI.		SMYTH.	
		Epoch.	Colours.	Epoch.	Colours.	Epoch.	Colours.
π Bootis	**517**	1836·5	*A. White* *B. White*	1844·4	*A. Yellow* *B. Less yellow*	1850·6	*A. White* *B. Creamy*
10 Hydræ	**519**	1831·5	*A. Pale orange* *B. Violet tint*	1846·4	*A. Yellow* *B. Yellow*	1851·4	*A. Deep yellow* *B. Reddish violet*
212 P. XIV. Libræ . .	**524**	1833·4	*A. Straw colour* *B. Yellow*	1846·3	*A. Orange* *B. Orange*	1851·4	*A. Yellow* *B. Dusky*
44 Bootis	**529**	1842·5	*A. Pale white* *B. Lucid grey*	1844·5	*A. Orange* *B. Orange*	1858 6	*A. Pale yellow* *B. Dusky*
δ Bootis	**537**	1835·5	*A. Pale yellow* *B. Light blue*	1844·5	*A. Gold yellow* *B. Yellowish azure*	1851·3	*A. Yellow* *B. Lilac*
μ' Bootis	**542**	1832·3	*A. Flushed white* *B. Greenish white*	1844·5	*A. Yellow* *B. Yellowish azure*	1850·6	*A. Yellowish* *B. Greenish white*
ζ Coronæ Borealis .	**549**	1842·6	*A. Bluish white* *B. Smalt blue*	1844·5	*A. White* *B. White*	1850·6	*A. Flushed white* *B. Bluish green*
51 Libræ	**558**	1842·5	*A. Bright white* *B. Pale yellow*	1846·4	*A. Orange* *B. Orange*	1850·6	*A. Creamy white* *B. Pale yellow*
β Scorpii	**559**	1835·4	*A. Pale white* *B. Lilac tinge*	1846·4	*A. Yellow* *B. Whitish*	1851·4	*A. Yellowish white* *B. Pale lilac*
κ' Herculis . . .	**560**	1835·4	*A. Light yellow* *B. Pale garnet*	1844·5	*A. Yellow* *B. Orange*	1851·3	*A. Pale yellow* *B. Reddish yellow*
ν Scorpii	**561**	1831·5	*A. Bright white* *B. Pale lilac*	1846·5	*A. Yellowish* *B. White*	1850·6	*A. Pale yellow* *B. Dusky*
σ Scorpii	**568**	1838·3	*A. Creamy white* *B. Lilac tint*	1846·5	*A. Yellow* *B. White*	1851·4	*A. Dusky white* *B. Plum colour*
236 P. XVI. Scorpii .	**593**	1833·4	*A. Yellowish white* *B. Pale green*	1846·5	*A. Yellow* *B. White*	1851·4	*A. Creamy white* *B. Greenish*
μ Draconis . . .	**602**	1839·5	*A. White* *B. White*	1844·5	*A. Yellow* *B. Azure*	1850·7	*A. White* *B. Pale white*
36 Ophiuchi . . .	**604**	1842·4	*A. Ruddy* *B. Pale yellow*	1846·5	*A. Orange yellow* *B. Orange yellow*	1851·4	*A. Ruddy tint* *B. Yellowish*
39 Ophiuchi . . .	**607**	1838·5	*A. Pale orange* *B. Blue*	1846·5	*A. Orange* *B. Yellow*	1851·4	*A. Pale orange* *B. Bluish*
ν Serpentis . . .	**610**	1832·6	*A. Pale sea-green* *B. Lilac*	1846·5	*A. Yellow* *B. Red*	1851·4	*A. Silvery tint* *B. Native copper*
ρ Herculis	**613**	1839·7	*A. Bluish white* *B. Pale emerald*	1844·4	*A. Yellow* *B. Deeper yellow*	1850·5	*A. Greyish* *B. Greenish*
53 Ophiuchi . . .	**618**	1836·5	*A. Bluish* *B. Bluish*	1844·5	*A. White* *B. Azure*	1850·5	*A. Greyish* *B. Pale blue*
95 Herculis	**631**	1833·8	*A. Greenish* *B. Cherry red*	1844·5	*A. Gold yellow* *B. Gold yellow*	1851·3	*A. Pale green* *B. Reddish*
70 Ophiuchi . . .	**633**	1842·5	*A. Pale topaz* *B. Violet*	1845·9	*A. Gold yellow* *B. Gold yellow*	1849·5	*A. Topaz yellow* *B. Purplish*
o Draconis . . .	**672**	1837·9	*A. Orange yellow* *B. Lilac*	1844·5	*A. Fine orange* *B. Copper colour*	1851·3	*A. Orange* *B. Lilac*
15 Aquilæ	**678**	1831·6	*A. White* *B. Lilac tint*	1846·5	*A. Reddish* *B. Red orange*	1851·4	*A. Yellowish white* *B. Red lilac*
28 Aquilæ	**690**	1831·4	*A. Pale white* *B. Deep blue*	1844·5	*A. White* *B. Yellow*	1851·4	*A. Dusky white* *B. Lilac blue*
β Cygni	**700**	1837·6	*A. Topaz yellow* *B. Sapphire blue*	1844·5	*A. Orange gold* *B. Azure*	1849·6	*A. Golden yellow* *B. Smalt blue*

Stars.	Cycle No.	SMYTH.		SESTINI.		SMYTH.	
		Epoch.	Colours.	Epoch.	Colours.	Epoch.	Colours.
ε Sagittæ	**704**	1833·8	*A. Pale white* *B. Light blue*	1844·5	*A. Yellow* *B. Bluish yellow*	1850·6	*A. Faint yellow* *B. Bluish*
54 Sagittarii . . .	**705**	1837·6	*A. Yellow* *B. Violet*	1846·5	*A. Orange* *B. White*	1850·7	*A. Yellow* *B. Pale lilac*
ζ Sagittæ	**718**	1838·6	*A. Silvery white* *B. Blue*	1844·5	*A. Yellowish white* *B. Azure*	1850·6	*A. Flushed white* *B. Cerulean blue*
56 Aquilæ	**722**	1834·6	*A. Deep yellow* *B. Pale blue*	1846·5	*A. Yellow* *B. Yellow*	1850·6	*A. Yellow* *B. Bluish*
κ Cephei	**743**	1838·8	*A. Bright white* *B. Smalt blue*	1844·6	*A. Yellowish* *B. Azure*	1851·3	*A. Pale yellow* *B. Blue*
γ Delphini . . .	**762**	1839·7	*A. Yellow* *B. Light emerald*	1844·5	*A. Orange* *B. Yellow*	1850·7	*A. Golden yellow* *B. Flushed grey*
ε Equulei	**770**	1838·8	*A. White* *B. Lilac*	1844·5	*A. Gold orange* *B. Azure*	1851·4	*A. Pale yellow* *B. Bluish lilac*
1 Pegasi	**782**	1833·9	*A. Pale orange* *B. Purplish*	1844·5	*A. Orange* *B. Azure*	1851·4	*A. Deep yellow* *B. Lilac blue*
β Cephei	**789**	1843·1	*A. White* *B. Blue*	1844·6	*A. White* *B. White*	1851·3	*A. Yellowish.* *B. Flushed blue*
3 Pegasi	**790**	1837·8	*A. White* *B. Pale blue*	1844·5	*A. White* *B. Yellow*	1850·5	*A. Flushed white* *B. Greyish*
ε Pegasi	**794**	1833·6	*A. Yellow* *B. Blue*	1844·5	*A. Gold yellow* *B Azure*	1851·4	*A. Bright yellow* *B. Blue lilac*
μ Cygni	**795**	1839·6	*A. White* *B. Blue*	1844·5	*A. Yellow* *B. More yellow*	1850·6	*A. White* *B. Pale blue*
29 Aquarii	**800**	1830·8	*A. Brilliant white* *B. White*	1846·5	*A. Red orange* *B. Same, lighter*	1852·7	*A. White* *B. Bluish*
ξ Cephei	**802**	1839·6	*A. Bluish* *B. Bluish*	1844·6	*A. White* *B. Yellowish*	1851·4	*A. Flushed* *B. Pale lilac*
ζ Aquarii	**813**	1842·6	*A. Very white* *B. White*	1845·8	*A. Orange yellow* *B. Pale yellow*	1849·2	*A. Flushed white* *B. Creamy*
δ Cephei	**815**	1837·7	*A. Orange tint* *B. Fine blue*	1844·6	*A. Orange* *B. Azure*	1849·2	*A. Deep yellow* *B. Cerulean blue*
τ1 Aquarii	**822**	1838·7	*A. White* *B. Pale garnet*	1845·8	*A. White* *B. Azure*	1849·2	*A. Pale white* *B. Flushed*
ψ1 Aquarii	**833**	1834·9	*A. Orange tint* *B. Sky blue*	1845·8	*A. Gold* *B. Azure*	1850·8	*A. Topaz yellow* *B. Cerulean blue*
94 Aquarii	**834**	1838·9	*A. Pale rose tint* *B. Light emerald*	1845·8	*A. Orange yellow* *B. Orange*	1850·8	*A. Orange tint* *B. Flushed blue*
101 P. XXIII. Cassiopeæ	**839**	1830·9	*A. Light yellow* *B. White*	1844·6	*A. White* *B. Yellowish*	1852·7	*A. Pale white* *B. Yellowish*
107 Aquarii	**844**	1832·8	*A. Bright white* *B. Blue*	1845·8	*A. Yellowish white* *B. Yellowish*	1850·7	*A. White* *B. Purplish*

All the differences in the above list are subject to several doubts, and many of the records have been noted without a very strict attention to the question. In the Cycle, the many disagreements are mentioned between the tints of stars as given by Sir William Herschel and myself; and the anomaly is partly accounted for by his peculiarity of vision, and partly by the composition of metal in his reflectors. But I am at a loss why refractors should differ so widely as here shown; and therefore hope the subject will be more closely pursued than it has hitherto been. We are aware that the notations independently made at various epochs will vary in term, though to the observer's mind they may mean nearly the same tint; still some of the differences mentioned by Signor Sestini in his interesting Memoir are singularly striking. He says—"Now, beginning with the companion of γ Andromedæ, we have Smyth emerald-green and Sestini white; but Herschel and Struve at another date call it azure. Moreover, observing it again after a lapse of two years, and four years after Smyth, I find it no longer white, but a strong blue!" And again—"Now see B (95) Herculis; according to Smyth one is greenish and the other red; but we think them both a golden yellow. A Ophiuchi, by Smyth, one ruddy and the other pale yellow; but we take them to be both orange. The contrary occurs in ι Bootis, the components of which by Smyth are both pale yellow; but we deem one to be orange and the other azure."

Anomalies.

Under the circumstances already alluded to, I am not at this stage disposed to theorise on the objects thus brought into juxta-position: and the colours of double-stars must be much more accurately assigned, and more ably experimented upon, before we can really admit that the nature and character of those suns actually do change in short periods. Sir David Brewster observes, that there can be no doubt that in the spectrum of every coloured star certain rays are wanting which exist in the solar spectrum; but we have no reason to believe that these absent rays are absorbed by any atmosphere through

Sir David Brewster's Experiment.

which they pass. And in recording the only observation perhaps yet made to analyse the light of the coloured stars, he says:—"In the orange-coloured star of the double-star ζ Herculis, I have observed that there are several defective bands. By applying a fine rock-salt prism, with the largest possible refracting angle, to this orange-star, as seen in Sir James South's great achromatic refractor, its spectrum had the annexed appearance (in the Campden Hill Journal), clearly showing that there was one defective band in the red space, and two or more in the blue space. Hence the colour of the star was orange, because there was a greater defect of blue than of red rays." This instance shows, that an approximation by instrumental means to the spectra of the brighter stars ought not to be despaired of; and that prospect should not admit of any relaxation in our present motive call.

Teneriffe Experiment.

In the year 1856, on my son's going to the island of Teneriffe to make his "Astronomical Experiment," it occurred to me that it would be a singularly fine opportunity to test sidereal polychromy; since it would be marked from a spot where some thousands of feet of the grossest portion of our atmosphere are eliminated.

This meritorious expedition, which "at once converted into an actual and successful fact a theoretical idea, long thought well of, but never previously carried into practice," may be recognized as the foundation of the branch so appropriately designated *Mountain Astronomy.* Indeed the advantages of elevated stations are sufficiently obvious to countenance the hope of reaping, thereby, results of the utmost importance to knowledge. Newton himself sounded a tocsin to this effect, in asserting that telescopes "cannot be so formed as to take away that confusion of rays which arises from the tremors of the atmosphere. The only remedy is a most serene and quiet air, such as may perhaps be found on the tops of the highest mountains above the grosser clouds." It was under a conviction of this truth—and having a personal knowledge of both sites—that I earnestly wished my friend Lassell to plant his huge equatoreal on Ætna, instead of Malta.

Such being the justifiable expectations of the problem, my son accordingly scrutinised the following stars from the "Cycle" for me; they having been chosen out of those objects which then happened to be in apparition from the mountain, during his interesting continuance upon it. The observations were made at two stations: the colours noted on the 29th of July and 4th of August were examined with the 5-foot Sheepshanks equatoreal, at Guajara, a height of 8870 feet; and those of September 4th, 5th, and 6th, were made with the Pattinson telescope of 7·25 inches aperture, and parallactic movement, at the Alta Vista, where the altitude is 11,000 feet. The following are his registered results, with comparisons from the "Cycle;" which last are of various dates from 1830 to 1843:—

OBJECTS.	TENERIFFE.		BEDFORD.	
	JULY 29th.			
α Herculis	A. *Cadmium yellow*	B. *Greenish*	A. *Orange*	B. *Emerald, or bluish green*
39 Ophiuchi	A. *Pale yellow*	B. *Faint blue*	A. *Pale orange*	B. *Blue*
5 [V?] Serpentis	A. *Pale yellow*	B. *Warm lilac*	A. *Pale yellow*	B. *Light grey*
ρ Herculis	A. *White*	B. *Bluish*	A. *Blueish white*	B. *Pale emerald*
95 Herculis	A. *White*	B. *White*	A. *Light apple green*	B. *Cherry red*
70 Ophiuchi	A. *Pale yellow*	B. *Greenish*	A. *Pale topaz*	B. *Violet*
α Lyræ	A. *White*	B. *Violet*	A. *Pale sapphire*	B. *Smalt blue*
	AUGUST 4th.			
α Herculis	A. *Cadmium yellow*	B. *Greenish*	*(as above)*	
95 Herculis	A. *and B. both yellow with tinge of bluish green.*		*(as above)*	
70 Ophiuchi	A. *Yellow*	B. *Warm green*	*(as above)*	
5 Aquilæ	A. *Pale yellow*	B. *Bluish.* C. *Blue*	A. *White*	B. *Lilac.* C. *Blue*
28 Aquilæ	A. *White*	B. *Blue*	A. *Pale white*	B. *Deep blue*
β Cygni	A. *Pale yellow*	B. *Blue*	A. *Topaz yellow*	B. *Sapphire blue*
186 Antinoi	A. *Yellow*	B. *Blue*	A. *Pale white*	B. *Sky blue*
	SEPTEMBER 4th.			
α Scorpii	A. *Coppery red*	B. *Blue*	A. *Fiery red*	B. *Pale*
α Herculis	A. *Orange*	B. *Greenish*	*(as above)*	

OBJECTS.	TENERIFFE.	BEDFORD.
	SEPTEMBER 4th—*continued.*	
ζ Sagittæ	A. *Yellow* . . B. *Blue*	A. *Silvery white* . B. *Blue*
α' Capricorni	A. *Yellow* . . a. *Blue*	A. *Pale yellow* . a. *Blue*
1 Pegasi	A. *Yellow* . . B. *Blue*	A. *Pale orange* . B. *Purplish*
β Cephei	A. *White* . . B. *Purple*	A. *White* . . B. *Blue*
3 Pegasi	A. *Whitish* . . B. *Warm grey*	A. *White* . . B. *Pale blue*
ζ Piscium	A. *Yellow* . . B. *Grey*	A. *Silver white* . B. *Pale grey*
γ Arietis	A. *White* . . B. *White*	A. *Bright white* . B. *Pale grey*
λ Arietis	A. *Pale yellow* . B. *Light lilac*	A. *Yellowish white* . B. *Blue*
α Piscium	A. *White* . . B. *White*	A. *Pale green* . B. *Blue*
	SEPTEMBER 5th.	
α Aquilæ	A. *Pale yellow* . B. *Grey*	A. *Pale yellow* . B. *Violet tint*
γ Delphini	A. *Cadmium yellow* B. *Greyish tinge*	A. *Yellow* . . B. *Light emerald*
τ' Aquarii	A. *Light yellow* . B. *Pale violet*	A. *White* . . B. *Pale garnet*
α Piscis Australis	A. *White* . . B. *Blue*	A. *Reddish* . . B. *Dusky blue*
ψ Aquarii	A. *Cadmium yellow* B. *Blue*	A. *Orange tint* . B. *Sky blue*
94 Aquarii	A. *Yellow* . . B. *Light warm lilac*	A. *Pale rose-tint* . B. *Light emerald*
101 Cassiopeæ	A. *Light yellow* . B. *Grey*, b. *blue*	A. *Light yellow* . B. *White*
———	C. *Blue* . . D. *Violet*	C. *Blue* . . D. *(not noticed)*
107 Aquarii	A. *Pale yellow* . B. *White*	A. *Bright white* . B. *Blue*
35 Piscium	A. *Yellow* . . B. *Pale violet*	A. *Pale white* . B *Violet tint*
113 Ceti	A. *Rich yellow* . B. *Warm grey*	A. *Cream yellow* . B. *Smalt blue*
γ Arietis	A. *Light yellow* . B *Light yellow*	*(as above)*
222 Arietis	A. *Grey* . . B. *Blue*	A. *Topaz yellow* . B. *Deep blue*
———	C. *Lilac* . . D. *Yellow*	C. *Lilac* . . D. *Pale blue*
α Piscium	A. *White* . . B. *White*	*(as above)*
γ Andromedæ	A. *Orange* . . B. *and* C. *Green*	A. *Orange* . . B. *Emerald green*
32 Eridani	A. *Orange* . . B. *Greenish*	A. *Topaz yellow* . B. *Sea green*
	SEPTEMBER 6th.	
σ Cassiopeæ	A. *Pale yellow* . B. *Light blue*	A. *Flushed white* . B. *Smalt blue*
35 Piscium	A. *Pale yellow* . B. *Pale lilac*	*(as above)*
113 Ceti	A. *Yellow* . . B. *Warm grey*	*(as above)*
146 Ceti	A. *Yellow* . . B. *Pale violet*	A. *Pale topaz* . B. *Violet tint*
η Cassiopeæ	A. *Yellow* . . B. *Indian red*	A. *Pale white* . B. *Purple*
65 Piscium	A. *White* . . B. *White*	A. *Pale yellow* . B. *Pale yellow*
ψ' Piscium	A. *White* . . B. *White*	A. *Silvery white* . B. *Silvery white*
ζ Piscium	A. *White* . . B. *Reddish*	*(as above)*

General discussion of the foregoing list.

In the preceding list there seems to be a very general similarity of eye-judgment between my son and myself; whence it would appear that the difference made by 11,000 feet of the lower atmosphere on the colours, is not so great as might have been anticipated. But the most striking and inexplicable difference in the comparison is that of 95 Herculis; for, in the observations at the Peak of Teneriffe, the tints of the two stars—though not quite the same at each examination—were judged to be common to both, and the impression was ratified by the evidence of some Spanish visitors at the astronomical aerie. Not a little taken aback, however, by the unexpected announcement—the more unexpected in consequence of the general agreement which existed throughout the list, even in some of the most delicate hues—I took the earliest opportunity of returning to the charge, when there was the A apple green and B cherry red, as recorded by me nearly a quarter of a century before! To avoid all suspicion of bias I invited my colleagues to the task, and soon received an answer from Mr. Dawes, saying—" On referring to my colour-estimations, I find that they agree very nearly with your own." Lord Wrottesley pronounced A to be greenish and B reddish in 1857·46; and at the same epoch Mr. Fletcher reported that A, to his eye, was light green and B pink; while to his brother H. A. Fletcher one was bluish green and the other orange, and to Mr. T. W. Carr A was either light blue or green, and B dull red. To add to the perplexity of the instance, Sestini saw them both golden yellow at Rome, in 1844; while his colleague, De Vico, in the memoirs of the " Osservatorio del Collegio Romano," in the same place and with the same instrument, dubs them " rossa e verde." Assuredly this is most passing strange, since the means in these cases were pretty equally powerful, and *chromatic personal equation*—or the faculty in a greater or less degree of appreciating differences of colour, cannot be entitled to consideration with the discrepancies of 95 Herculis.

Chromatic difficulties.

Under our present practice, various difficulties are presented, for the designation of hues uttered in mere parlance by several persons often means the same tints in different words, and these will not always quadrate with the chromatic language of photologists. The wonder, however, is not so great that, without a due nomenclature, we should differ so much from each other, and even from ourselves at different dates, but rather that, with such an unorganized practice, so many instances should coincide relatively. In many cases the difference of colour in the components of a double star are real; but when they are merely complementary, the fainter of the two may possibly be a white star which appears to have the colour complementary to that of its more brilliant companion. This is in consequence of a well-understood law of vision, by which the retina of the eye being excited by light of a particular colour, is rendered insensible to less intense light of the same hue,—so that the complement of the real light of the fainter star finds the retina more sensible to it, than to the ray which is identical in colour with the brighter star; and the impression of the complementary tint accordingly prevails. But the accurate perception of the colour of a celestial body often depends as much on the condition of the eye when the object is seen, as upon the object itself; and possibly the achromatism of the object-glass, which, being adapted to the solar spectrum may not be suitable to the spectrum of a star, ought to be taken into account; as well as a nice adjustment of the eye-piece, to lead to a discrimination between real and illusive appearances. The powers of colours in contrasting with each other, agree with their correlative powers of light and shade; and such are to be distinguished from their powers individually on the eye, which are those of light alone. It may assist the memory of the inexperienced observer, to remind him that the primary colours and their complementaries are in these relations—

RED . . . GREEN
BLUE . . . ORANGE
YELLOW . . VIOLET

and from these a scale may be readily drawn up of the

subsidiary tints and their opposites (the male and female lights of Milton*), through all the twistings of Iris: and if he will bear in mind the laws of harmonious alliance and contrast of colour—that yellow is of all hues the nearest related to light, and its complementary violet or purple to darkness—that red is the most exciting and positive of all colours, and green the most grateful—that blue is the coldest of all hues, whilst orange is the warmest—much of the apparent mystery of harmonizing the multiplied tints of primary, secondary, and tertiary colours, will be readily accounted for.

Vibrations of light.

In the present incertitude, it is suggested that variations in colour may be owing to variations in stellar velocity; but in this case would there not also be as palpable a variation in brightness? If it shall be found that the tints actually vary, the comparative magnitude should also be carefully noted, to establish whether a variability in brightness accompanies the changes of colour. Sestini, however, does not view the matter in this light: he holds that the undulations of each colour arrive in succession to our eyes, and that therefore at last, when they have all reached us, the result will appear white. In arguing the circumstances necessary for the case—as the strength of vibrations with their number and velocity in a given time—he cites Huyghens, Euler, Young, Fresnel, and Arago. Quoting Herschel's data, he observes, that five hundred and thirty-six billions of vibrations cause us to see yellow, while six hundred and twenty-five billions exceed the number that shews blue: that is, when the tangential celerity of the moving star in relation to its companion, comes at its maximum to equal one-thir-

* The notion of male light being imparted by the Sun, and female light by the Moon, is as old as the hills. Pliny, in his CYCLOPÆDIA (*lib. ii. cap.* 100 *and* 101,) mentions it as a condition "which we have been taught;" and he details the influences of the masculine and feminine stars. Here, probably, Milton imbibed the hint to which I alluded in the Cycle (I. page 301) —

> "Other suns, perhaps,
> With their attendant moons, thou wilt descry,
> Communicating *male* and *female* light."

teenth of that of light. Its green colour will change insensibly into yellow on increasing its distance, and then, receding through the same steps, it will again become green; beyond which, as it approaches the eye, it will become a full blue; finally, in the inverse order, it will turn to green, and so on. But this explanation is not admissible, as may be readily shown: for instance, if we accelerate the velocity of the star to one-fifth of that of light, we shall have the number of vibrations corresponding to red = four hundred and eighty-one billions, and seven hundred and twenty-one billions, which exceeds that of violet. In this supposition, the green star when furthest from its companion will become red, and when approaching it must be of an intensely strong violet tinge; after which, owing to its circular orbit, it will in receding again become green, thus passing through all the colours of the spectrum. These are the ratios—

$$536 : 625 :: 1-\tfrac{1}{13} : 1+\tfrac{1}{13} :: 12 : 14 :: 6 : 7.$$
$$481 : 721 :: 1-\tfrac{1}{5} : 1+\tfrac{1}{5} :: 4 : 6 :: 2 : 3.$$

Decision enjoined.

Admitting these and the like grounds, as the laws of new stars and binary systems may be somewhat elucidated thereby, I strongly recommend repeated examinations of the brightness and colours of stars to the well-equipped amateur, who is also happily possessed of a good eye, perseverance, and accurate notation. But even thus prepared, I would advise him, before entering upon the undertaking, to study well the third chapter of the great work of my highly-esteemed friend Sir John Herschel, on the Uranography of the Southern Hemisphere: it treats of Astrometry, or the numerical expression of the apparent magnitudes of the stars. In a more advanced state of this question the measurement of brightness should always accompany that of colours, since a change in the one might possibly produce variation in the other: and who can say that numerical measures may not be made with such extreme precision hereafter, that the distance of stars thereby may be given? The observer must not however be unnerved by the difficulties, some of them apparently insuperable, which beset the inquiry: nor by the philosopher's assertion that "nothing short of a separate and independent estimation of the total amount of the red, the yellow, and the blue rays in the spectrum of each star would suffice for the resolution of the problem of astrometry, in the strictness of its numerical acceptation; and this the actual state of optical science leaves us destitute of the means even of attempting, with the slightest prospect of success." This might indeed be a damper to our argument, so far at least as stars differing in colour are concerned; but perseverance in a good cause has often been rewarded with marvellous accomplishments,—and it is well to remember that

> By many blows that work is done,
> Which cannot be achieved with one.

On stellar velocity.

These remarks will hardly be impinged upon in practice, by taking one objection to the facts upon which Sestini's theory is founded, namely, the velocity of the stars; since, in the present day—even admitting proper motions and translations in space to their fullest extent—it is not necessary to consider the possible rate of sidereal movements as capable of bearing any sensible ratio to the speed of light. In citing the case of the orbital velocity of the companion of a double star, he should have applied it to *a* Centauri, an object of which we know all the elements, its distance from us and from each other in miles, the mass of the components as compared to our Sun, their quantity of light as compared to the same, and the periodic time;—all these we know to a greater degree of confidence than those of any other similar body. Now the theory fails upon this test; for the mean orbital velocity of the companion may be assumed as 2·5 miles per second, while Sestini's limits of $\frac{1}{13}$ and $\frac{1}{5}$ of the velocity of light would make it fifteen thousand and thirty-eight thousand miles, in the same time. The velocity of light assumed here is, however, it must always be remembered, that of the Sun; that determined by direct observations of the solar orb itself, or by the eclipses of Jupiter's satellites, whose reflections still give us solar light, and traversing the same medium, whatever it be, filling the planetary spaces. But we may reasonably expect, and, indeed, the experiments detailed above, on the spectra of different stars, appear to indicate, if not actually to prove, that the light of some of the stars is absolutely of a distinct nature, and radically of a different composition, to that of the Sun; while the media also which the rays have to pass through may be of a kind unknown in any part of the whole of our planetary circles, and of a nature the peculiarities of which we are at present profoundly ignorant of.

Evidently, therefore, when the speed of transmission of the stellar rays comes into play, we may have to deal with velocities very different to that on which our correction for aberration—which depends upon solar light—is founded: the speed of transmission of which element, the velocity of electric light, and the speed of sidereal light, appear to depend upon, or be affected by, different causes. Granting that, however, and to the widest extent; extending even the somewhat doubtful experiments which have been made on the velocity of electric light, as compared with the solar, and on the transmission of ordinary light through air and through the denser medium of water; still there is nothing as yet to show, that we are likely to meet with any kind of light moving at so slow a rate, as to bear the proportion (which Sestini's theory requires) to the actual speed at which any star has been found to move.

There is, however, another way in which the peculiar habitude of rays of light may produce a difference of colour in a star, and make it even run through the whole of the colours of the spectrum from one end to the other and back again, in a greater or less space of time according to the particular circumstances of the case. This will occur if the different coloured rays of which the white beam is composed undergo intrinsically in themselves, or by reason of the nature of the medium which they traverse, any difference in the velocity of their transmission.

Emission and undulation.

According to the Newtonian doctrine of "emission," the

separate colours are actually produced by different degrees of velocity: and he concluded—from experiment—that the transparent parts of bodies, according to their several sizes, reflect rays of one colour and transmit those of another. But, according to the " undulating " theory, which has since been shown by Young and Fresnel to be far more probable than the other, if not really to be the true theory, the various tints are produced by means of undulations of different lengths; and the physicists have even been able to measure the comparative extent of these minute waves, or undulations, and have assigned them decimal proportions as follows:

	Parts of an Inch.
Red	0·00002582*
Orange . . .	0·00002319
Yellow . . .	0·00002270
Green . . .	0·00002073
Blue	0·00001912
Indigo . . .	0·00001692
Violet . . .	0·00001572

Now, though this by itself may say nothing with respect to the rapidity with which each undulation may be transmitted, it renders the probability of such a difference extremely great; and, though that difference be so very small that there is no hope of ever being able to make it manifest in any scientific apparatus of even the most delicate description, yet, on account of the vast remoteness of the stars, the effect may become at length very sensible. For, although the

* These lengths of an undulation lead to the astounding inference, that, on viewing a red object, the membrane of the eye trembles at the rate of 480,000000,000000 of times in every two beats of a seconds' pendulum! The researches and discoveries of Huyghens, Young, Malus, Fresnel, Arago, Poisson, Airy, Wheatstone, and others, have rendered the hypothesis of an undulatory propagation of light almost a demonstrated truth. "It is a theory," says Herschel, "which, if not founded in nature, is certainly one of the happiest fictions that the genius of man has yet invented to group together natural phenomena, as well as the most fortunate in the support it has received from all classes of new phenomena, which at their discovery seemed in irreconcileable opposition to it. It is, in fact, in all its applications and details, one succession of *felicities;* inasmuch that we may almost be induced to say, if it be not true, it deserves to be."

difference in the rate of propagation by the waves of each ray may be the smallest conceiveable quantity, yet, if that different rate be kept up during the whole of the one thousand years that we suspect must be occupied by the light of some of the stars in reaching us, notwithstanding that it may travel on the average one hundred and ninety-two thousand miles in a second, it is manifest that, after continuing to grow during so great a length of time, a very decided effect may at last be produced.

Assumed instance.

If a new star suddenly appears in any part of the sky, the rays of light immediately travel off to announce the fact everywhere, and to us amongst the number of other orbs; and it matters not whether the light consist in the emission of particles, or the propagation of waves of different orders, as many of Arago's "couriers" as there are different colours in the spectrum are sent off with the intelligence; and, if one is able to accomplish the great intervening distance between the star and us in a less space of time than the others, and so arrive before them, we shall see the star of that colour first, say violet. In that case the next to arrive would be the yellow, and then arriving and mixing with the blue, already come, would make the star change from pure blue to green; while the positive red, arriving last of all, and joining themselves to the existing green, would at length make the star appear white; and, if it preserved the same lustre, it would ever after continue white.

Reminder.

But be it recollected that, in the ideas evoked by the discrepancies of colour-estimates, I am only throwing out suggestions, not advocating an hypothesis; still it must be admitted that variations of colour ought to accompany variations of brightness, though such variation of hue has not hitherto been detected in some stars that notably vary in splendour. With this confession, we will proceed in the vision thus conjured up, and return to the celestial body in white; only reminding the reader, that little is correctly understood of light in its causes and principles of existence, and that Bacon has told us—*rectè scire est per causas scire.**

Argument resumed.

If the above-cited star be shown for only an instant of time as an electric spark, then we might see it varying through each of the different colours, blue, yellow, and red, separately and distinctly. Allowing that, for example, the blue ray was to traverse the space between the star and ourselves in three years, the yellow in three years and one week, and the red in three years and two weeks, and supposing the above to apply only to the central portion of each coloured ray, which should gradually vary with filaments of different velocities so as to join insensibly with those of the neighbouring one; then, three years after the striking of this stellar spark, we should see a blue star appear in the sky, and last for one week; then the star would appear yellow during another week, and red during another; after which it would be lost altogether. Or if there be actual separations between the different colours, as is more than hinted at by the discovery of the black bands in the spectrum, then the star, after appearing of one colour, might even disappear for a time before the next colour began to arrive.

Again, if a star which has existed for ages be on a sudden extinguished, the rays last emitted will be the couriers to announce the fact; and, supposing the star to have been white, three years afterwards (in the above particular example), the last of the blue rays having arrived before the last of the others, the blue will be deficient in the star, and from white it will become orange; after a week all the yellow ones will have come in, and the star will be red; and, when the final rays of this colour have arrived, it will totally disappear.

* In a letter of March 10th, 1860, the Master of Trinity, Dr. Whewell, says—"If I was writing a review of your splendid volume (*Speculum Hartwellianum*), I should of course try to find some fault in it by way of showing my acuteness: and I should say—'*At page* 324, *we read Bacon has told us — rectè scire est per causas scire.*' We may remark that it was Aristotle who said this in Greek, and his followers in Latin. We may add that it is not a maxim which has done much good in science, for the first step is to learn the laws of the phenomena. The cause afterwards if we can: but, if we cannot, we have still learned something."

But if the star shines permanently, and has so shone from time immemorial, then, whatever might be the difference of time elapsing between the blue and red rays shot from the star at the same instant reaching us, we should see the star white; for blue and yellow and red rays of different dates of emission would all be reaching our eyes together.

Practical exemplification

This case can be exemplified by looking through a prism at a white surface of unlimited extent and equal brightness, when it will be seen as white as before; for the multitudinous spectra formed by all the component points of the whole surface overlaying each other, the red of one coming to the blue and yellow of others, will form white light as completely as if the three colours of one point be concentrated together again. Here was Goethe's error: he gazed at a white wall through a prism, and, finding it white still, kicked at Newton's theory to produce an absurd one of his own. But had he looked at the edges of the wall—which is a similar case to the birth or death of a star—he would have seen the blue half of the spectrum on one side, and the red on another: everything, in fact, with a sensible breadth will have coloured borders, blue on one side and red on the other. If one part of the wall, however, be brighter than another, the strong blue of that portion thrown on to the fainter red of another, will give that a bluish tinge, and *vice versa;* and so with the stars; if their brightness should alter, or, in the common though singularly erroneous parlance, their magnitudes vary, the strong blue of a bright epoch arriving with faint red of a dull period, will make blue appear to us as the predominating colour; will cause indeed the star's light to appear decidedly blue at one time, and, *mutatis mutandis*, red at another, although all the while the star's colour may not really have altered at all; but may have been really, and would have appeared to observers close by, as white as ever, varying only in quantity and not in quality. Real alterations in colour may doubtless occur, yet evidently may also often be only consequences of alterations in brightness, which may be brought about by many regular and periodical phenomena, and certainly do not require the introduction of any such startling reason as the conflagration that was lugged in to explain the tints through which the variable star of 1572 passed, as it gradually died out of the sky, where it had so suddenly appeared a few months previously. Of this, at least, we may be certain, that there are periodical variations in the brightness of the stars, and that some alteration of colour should thereby be produced; but whether to a sensible extent or not, is only to be determined by experiment. β Persei has been selected by Arago as a favourable instance for testing this matter by observation, because it changes so very rapidly in brightness in a short space of time; but, though he did not succeed in detecting any alteration of colour we must not despair; for, while on the one hand his means of determining the colour seem to have had no sensible degree of exactness, it is easily possible to assume such a difference of velocities for the various coloured rays of the star, and such a distance for them to traverse, as should completely annihilate the expected good effect of the quickness and frequent recurrence of the changes in this particular star. Many other stars might indeed be picked out where the natural circumstances are more promising, while further steps towards the perfection of the means of observation, would allow of many more still being made subservient to the inquiry.

How to observe.

The failures made heretofore may therefore be regarded in the same light as those in the olden inquiry of finding the parallax of the fixed stars, viz. not as reasons for leaving off, but for trying again more energetically, more extensively, and with more accurate means than before; and, although I may not be prepared just at present to describe any perfectly satisfactory method of observation, still, as some amateurs desirous of pursuing the subject may like to see such hints as my experience has incidentally given rise to, presented in some rather more practical form, I have thrown them together as follows:

In any method of determining colours of stars, three possible sources of error have to be met: 1. The state of the atmosphere generally at the time in altering the colour of all the stars above the horizon; 2. The effect of altitude in varying on different stars the apparent colour produced by the atmosphere; and 3. The effect on the eye of the necessary quantity of some sort or other of artificial light, for the purpose of writing down or examining the dimensions of the instrument, the face of the clock, &c. &c.

First source.

The first can only be eliminated by extensive observation of a number of stars, especially circumpolar ones, all through the year. Although the colours of some stars may vary in a small number of months, weeks, or even days, the mean of them all may be considered to be safely depended on for a tolerably constant quantity; and each star should be examined and tested for its colour every night, by comparison with the mean of all the rest; and where any decided variation appears to be going on through the year, that star should at once be excluded from the standard list, and its difference from the mean of the others stated as its colour for each night's observation.

Second source.

The second source of error is to be met by observations of the same star through a large part of its path from rising to culminating, or a number of stars of known colour at various altitudes, combined with a correction something similar to that for refraction, as varying in a proportion not far from the tangent of the zenith distance; and which would consequently require the altitude of every body observed to be carefully noted, as a decidedly necessary element in reducing the observations.

Low stars, however, should be eschewed, and each observer should confine himself as far as possible to his zenith stars; for, in addition to the low ones being so much fainter to him, than to one to whom they are vertical, and in addition to the colouring and absorbing effect of the atmosphere increasing so excessively low down on the horizon, the envelope acts so strongly there as a prism, that, combined with the bad definition prevailing, I have sometimes seen a large star of a really white colour appear like a blue and red handkerchief fluttering in the wind: the blue and red about as intense and decided as they could well be. This shows the extreme importance of noting not only the altitude of the star, which determines also the degree of prismatic effect, but of distinguishing in the observation any difference between the upper and lower parts of the star. In the Sun and Moon, bodies of very sensible breadth, this effect is not so evident; the surface will still be white or coloured uniformly by the atmosphere, and the upper and lower borders will alone show the prismatic colours, half on one edge and the other half on the other, as in the case of the white wall mentioned above; but the star under discussion, being merely a point of light, is wholly acted on, and exhibits as complete a spectrum as could be contrived without any of the white or self-compensating intermediate portion.

Combined with this is the colouring effect of the object-glass, and any deficiency in its achromaticity; though these, being nearly the same on all the stars, will not affect the difference observed: yet the latter quality of the eye-piece will be of more consequence, unless the star be brought very rigorously into the centre of the field of view, and kept there the whole time that it is under observation. A well achromatized eye-piece should be specially used, and its assigned magnifying power always recorded.

Third source.

The third difficulty may be best counteracted by using one eye for the field of the telescope, and the other for writing down, &c.; having the artificial lights used for these purposes as faint and making them as white as possible, with various other little practical details which will best occur to each observer.

We then come to the grand difficulty: viz. the manner in which the colour is to be determined; the methods are two: first, by the senses; second, by instrumental means. The first is that which has been employed hitherto, and will doubtless still be the only method employed for a considerable time by amateurs; and, though so very vague, yet may—by the education and the practice of the senses, combined with the corrections above considered—be carried to considerable perfection: still the education must be much more systematic, and the practice much more constant, than they have hitherto been. Nor will the pursuit be altogether unfruitful, even if it only relieves science by thereby proving a negative; but to the zealous aspirant there is a hopeful guerdon, because much of the theory of the universe may be finally revealed by this elegant and difficult element.

Standard reference required.

Some certain standard of colours must be kept and constantly referred to: the colours of precious stones have been used for this purpose; but, though very proper in one point of view, as being by their brightness more comparable to stars than ordinary pigments are, yet astronomers in general have not much acquaintance with anything so valuable and costly; and, if they had, would find that the colour of each star is not certainly to be defined by the jewel, *i. e.* that under the same name many different colours may be found; and different observers will therefore be giving the same name to stars not resembling each other; in addition to which there is not a sufficient range of colours amongst the precious stones to meet all the cases which occur in nature in the heavens, and they neither admit of being mixed, to form varieties of colour, nor of being modified, to show gradations in their own colour; a most important defect. These qualities, however, are possessed by the water colours of the present day; the greater part of them are very permanent, and the others, which are not so, are capable of being prepared fresh and fresh; the number of colours moreover is great, the combinations that may be formed of them almost endless; and gradations of each may be made, from nearly white to all but black. Not only must a scale of them be had in possession, and frequently referred to, but it must be made and remade by the observer, as a mode of impressing the colours on his memory; and, unless he can carry them in his mind, he need not attempt the chromatic observation of stars; for as he cannot see the star and his scale of colour at the same instant, and side by side, the estimate of the star depends entirely on the accidentals of memory.*

A word to the Tyro.

Not to be too dogmatical, however, with the willing neophyte in his outset, we may observe, that, though the aptitude thus recommended may be troublesome to attain, it is approachable by opening trenches; and nature has kindly provided many individuals with the requisites for receiving delicate impressions by the senses. Now the source of colours is acknowledged to be light: each primary tone being surrounded by its harmonizing secondary, which is again bordered by its tertiary; and the perceptive faculty of at once distinguishing their several shades is an endowment of the most pleasing power, whether exerted in a passion for stars, flowers, and mundane finery—or in contributing to render the painter's art impressive to the imagination, and delightful to the eye. While therefore a spectator is able to enjoy the sensitive perception of the varied gradations of hues and tints, he need not involve himself in the *vexata questio* as to colours being material or not—whether entities or individualities. That they are not yet really reducible to a single principle, is no reason why they should not be used most comprehensively, and every advance must be duly encouraged for the results that may ensue.

The Hartwell experiment.

We tried an experiment on chromatic personal equation, in its simplest form, at Hartwell, on a fine evening, the second of July, 1829. Having prepared a stone pedestal in front of the south portico of the house, on which was placed a Gregorian telescope of 5½ inches aperture, a party of visitors, consisting of six ladies and five gentlemen, were invited to gaze upon the double-star Cor Caroli; and they

* Chromatography is not so near perfection as the power of the eye and state of art would lead us to suppose it to be; but it is hoped that Mr. Chevreul's beautiful work on Colours, which has appeared since the above was printed, will yield a useful standard of tints for astrometry, as well as for manufactures, so as to afford an easy and ready reference.

were each to tell me—*sotto voce* to prevent bias—what they deemed the respective colours of the components to be. The first who stepped out by request, was my good friend the late Rev. Mr. Pawsey—more addicted to heraldry than to astronomy—who, after a very momentary snatch, flatly declared that he "could make out nothing particular:" but the other spectators were a little more attentive to the plan proposed, and their respective impressions were thus noted down in the large Hartwell Album:—

Observer	A	B
Mrs. Tyndale	A. *Pale white.*	B. *Violet tint.*
Mrs. Rush	A. *Yellowish cast.*	B. *Deadish purple.*
Miss Honor	A. *Yellowish.*	B. *Lilac.*
Miss Charlotte	A. *Light Dingy yellow.*	B. *Lilac.*
Miss Emily	A. *White.*	B. *Plum colour.*
Miss Mary Anne	A. *Paleish yellow.*	B. *Blue.*
Mr. Rose	A. *Cream colour.*	B. *Violet cream.*
Mr. B. Smith	A. *Pale blue.*	B. *Darker blue.*
Dr. Lee	A. *Whitish.*	B. *Light purple.*
Capt. Smyth	A. *White.*	B. *Plum-colour purple.*

Now, whatever may be said about instrumental means, tendency of metallic mirrors, weather influence, atmospheric light, or the object's position as to meridian, it is clearly obvious that every condition was common to the whole party, and we doubtlessly all meant the same hues. It must be admitted, however, that the stars were new to most of the spectators, and, though some of the eyes were surpassingly bright, they had never been drilled among the celestials. Further observation, with an achromatic instrument, led me to record Cor Caroli in the Cycle for 1837, A flushed white, and B pale lilac; but, as Sestini found them to be yellow and blue in 1844, I again probed them in 1850, when A struck me as full white and B very pale, but slightly ruddy under that paleness. From the lightness of the tints, this object offers less distinctness than deeper-coloured stars; insomuch that in 1830 Herschel said—"With all attention I could perceive no contrast of colours;" yet, at my last inspection in 1855, three observers were unanimous that A appeared to be a pale reddish white, and B lilac, under a magnifying power of 240, and a fair sky. All this shows the urgent necessity of a chromatic scale being drawn up for general adoption; and that, as yet, we are only on the threshold of a very beautiful department of knowledge.

Mistaken notion.

Many persons may think that a mere glance at colours is enough to impress them at once on the memory, and that, without any practice at that sort of remembrance, they can keep any tint in their mind for a length of time; but a more erroneous idea was never entertained. To these unhappy persons greens are greens, and blues blues; for they have never entered the magical region of colours, whereby a whole world of intellectual enjoyment is for ever closed against them. Bring them to the proof of their boasted powers; show them any portion of a landscape; and then place colours before them, and make them put down the various tints from memory, but this a week or two after the scene was witnessed. If hardy enough to attempt the task, every one of their tints will be found in error, and they will only put down one where nature had fifty. Even the painters confess, that, though colour may be a low branch of their art, yet it is the most difficult. Just look at the walls of the Royal Academy and see how rarely is a good colourist to be met with, and, when he is, how the initiated will gloat over the matchless and magic variety and mellowness of tints, while the uninitiated can see barely more than one, and that to them not noticeably different from the world of common-places beside it. Only look, too, at the characteristics of those painters who *draw* from nature, but do not *colour* also from her; who make their sketches in the open air with pencil or sepia, and fancy colouring to be so simple and so easily remembered, that they may do that afterwards comfortably at their home. Such works are detected wherever they are seen, by the poverty of tints, and by the uniformly monotonous colours that are always employed in the same manner. The human mind cannot invent to any extent, but can merely put together in a novel manner materials collected from the external world. Hence such materials in colouring can only be impressed on the memory by actual pains-taking and laborious copying and working from nature, by making the tints and applying them in imitation of her. By such training, this branch of memory may be strengthened as well as any other; for we find that the works of artists who adopt this method are always superior in their colouring to those of others, even when they paint from retentive memory or imagination. And one of the best colourists that we have ever had in landscape-painting was so impressed with the importance of cultivating the memory in this manner, that he used, even in the days of his prosperity and highest prices of his works, to spend much time in the open air making studies in oil, and then, as soon as they were made, tore them up; so that, as the followers of Cortez saw the necessity of conquering when their commander burnt the ships in which they might have made an inglorious retreat, and exerted themselves accordingly,—in the same way, not being able to refer, when painting a picture at home, to the sketch made in the open air, he felt himself necessarily obliged to tax his powers of memory, and make them exert themselves to the very utmost.

Signor G. Lusieri.

In the early part of this century, it was my good fortune while in Athens to make the acquaintance of the late Signor Gianbattista Lusieri, the eminent landscape-painter engaged by Lord Elgin to work in Greece. This philosophical artist showed me a series of views, proving his gradual improvement through twenty years, by making Nature his model throughout; and he restricted himself, moreover, to the same hour of the day for colour; so that some pictures which he was unable to complete before a change of weather, he reserved till the same season of the following year. Hence in great measure arose the perfection of his pencil.

Spectral lines.

The second, or instrumental, method of determining colour, need not be entered upon at much length here, as mere amateurs are not very likely to practise it; and would be working at a great disadvantage compared with any instrument in a public observatory specially devoted to this object. Brightness is everything under such appliance, and this must be commanded both by elevating the telescope into a high region of the atmosphere, and by adopting the largest possible size of aperture; for, not only must photometrical determinations of the lucidity of different sections of the spectra of stars formed by prisms be made, but the black lines in the spectra of each star must also be carefully examined into, as all the transcendent intelligibility of modern optics depends on them. Still the task appears to be equally prolix and toilsome; and, moreover, spectrum examinations are much more fatiguing to the eye than ordinary telescopic work, while it is only on the finest nights that the lines in the stellar spectra are steady enough for measurement. Indeed the difficulties of observation are now so complex, that the *complete* scrutiny of the spectrum of a single star may probably be the work of some years.*

Wollaston and Frauenhofer.

It is under no depressing view of the effects of progress that I perceive, however hopeful anticipations may be indulged in, the alliance between Astronomy and Chemistry is not yet definitively ratified; so that at present it is rather perplexing to speculate upon the ultimate results of the connection. Suffice it here to note, it is to my late eminent friend Wollaston that we owe the discovery of the existence of a peculiarity in solar light, which revealed the deficient rays causing the black lines of the spectrum: and this was followed quite independently by the marvellous measures of them by Frauenhofer of Munich, whose accurate determinations of their distances now form standard points of reference for apportioning the refractive powers of different media on the rays of light. These operations opened out a way for the masterly metallic investigations of Kirchhoff and Bunsen. Such skilful and well-conducted re-

* Notwithstanding this sombre view, there are rays of hope on the horizon. Already has the Astronomer-Royal directed his powerful abilities to the subject-matter; and he has suggested an apparatus for facilitating observations of the spectra of stars. This is described in the twenty-third volume of the Astronomical Society's Notices, pages 188—191.

searches are pregnant with the best augury; and the elaboration of successive variations of them must infallibly aid our future advance in the knowledge of STELLAR PHYSICS.

Advantages of instrumental method.

There is still, however, much to achieve before full reliance can be indulged in, as well respecting drilling the eye as the manufacture and use of its apparatus, and the subsequent delicate manipulation. When genius and perseverance shall have brought these conditions to bear in concentrated practice, then will this instrumental method of reducing colour to brightness and place prove invaluable; because—in addition to the exactness of the numerical determination of which it will then be capable—it would further overcome a most serious source of error—one which has barely been touched upon in all that has gone before in this our lucubration, and may affect to its fullest extent the method of the "senses," namely, chromatic personal equation. In fine, it seems destined to become another means of augmenting the debt which pure Astronomy owes to the powers of practical observation—which is the basis of all we know in the argument.

Proposed diagram of colours.

The instrumental desiderata above alluded to are mostly, as yet, irreducible to general practice, especially upon minute or dim objects; and, therefore, until certain difficulties shall be overcome, the amateur may still render good service to the cause, by noticing the stellar hues according to the scheme of colours submitted to him on the chromatic diagram hereunto appended. Where the mental impression is not quite adequately represented by these tints, it can be modified by an expressive adjective, as blueish green, brownish yellow, and the like; but, in pronouncing upon very delicate distinctions, the observer must keep in mind that all colours are accompanied by their accidental or opposite tinctures, or those which depend on the affections of the eye, rather than on the light itself—and, when the direct and accidental colours are of the same intensity, the accidental is then called complementary, because completing the series with the direct hue.

The very numerous shades from white to pale yellow are so

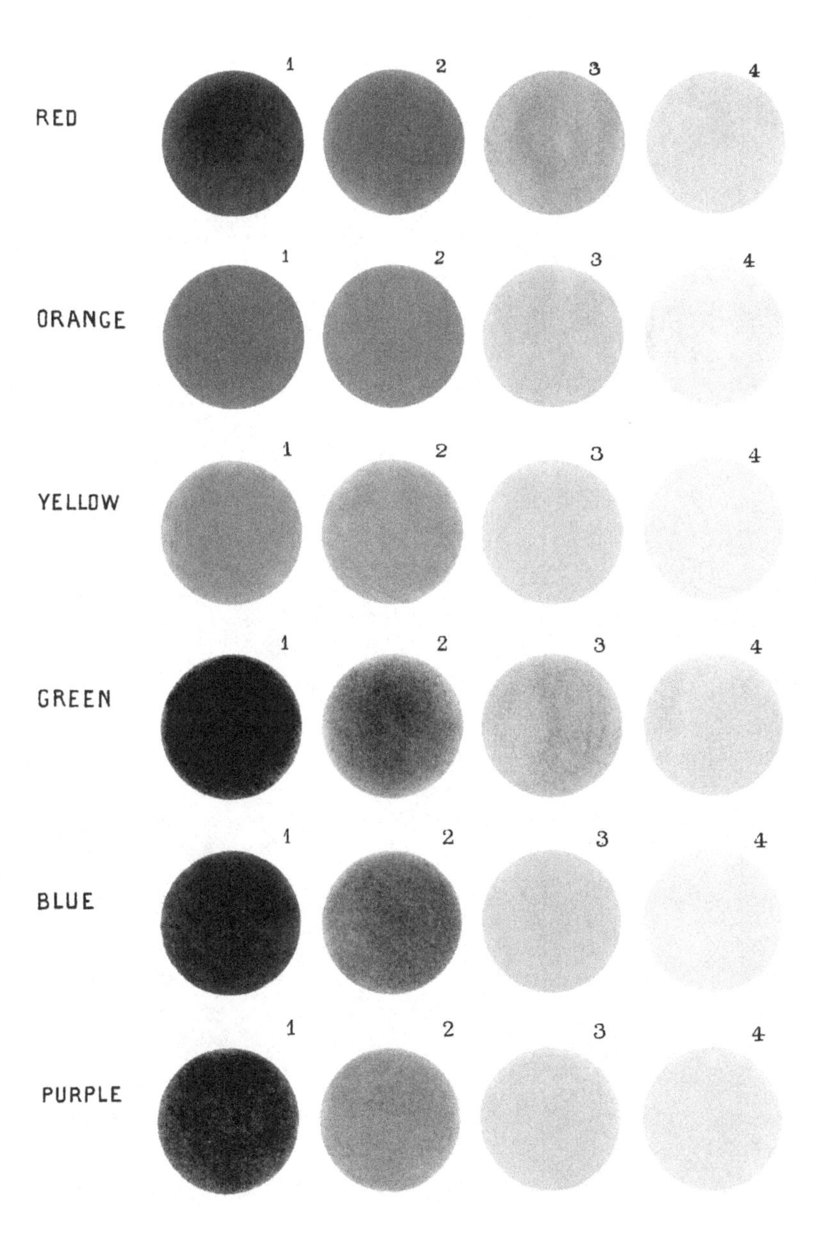
RED
ORANGE
YELLOW
GREEN
BLUE
PURPLE
1
2
3
4

unfit for representation and lamplight reference, that they are omitted in the annexed form; but the careful observer may readily estimate the intensity of almost colourless bodies according to the following order—Creamy white 1, Silvery white 2, Pearl white 3, and Pale white 4.

Notwithstanding the amount of obstacles that I have mustered up, they are not insuperable to resolution; for, as in mundane affairs, the word *impossible* ought to meet with no encouragement in scientific emprise. A specimen of the advantages of a more exact method than that which has hitherto obtained among us, may be here given, as showing that the glow ascribed can readily be consulted, which the mere guessing cannot. Indeed, if memory is to be trusted so far, some notations on the old system may be contrasted with our proposed plan, thus—

STAR.	CYCLE.	DIAGRAM.
51 Piscium . .	A. *Pearl White* B. *Lilac tint*	A. *White* 3 B. *Purple* 3
γ Andromedæ .	A. *Orange colour* B. *Emerald green*	A. *Orange* 2 B. *Bluish green* 3
40 Persei . .	A. *Pale white* B. *Ash-coloured*	A. *White* B. *Blue* 4
225 P. Aurigæ .	A. *Creamy white* B. *Pale grey*	A. *White* 1 B. *Blue* 4
δ Hydræ . .	A. *Light topaz* B. *Livid*	A. *Yellow* 4 B. *Purple* 3
17 Virginis . .	A. *Light rose-tint* B. *Dusky red*	A. *Red* 4 B. *Red* 3
ι Bootis . .	A. *Pale orange* B. *Sea-green*	A. *Orange* 3 B. *Green* 4
α Herculis . .	A. *Orange* B. *Emerald, or bluish green*	A. *Yellow* 2 B. *Blue* 3

Instance of β Cygni.

By the adoption of the new and firmer method of registry, many of the differences now on record may turn out to be more apparent than real, and will probably disappear before the organised process of using an accepted chromatic scale with educated eyes. We have alluded to the otherwise complicated embarrassments and uncertainties incident to this kind of observation; yet still there is basis enough to inspire the warmest hope for exertion. The beautiful pair of stars forming β CYGNI may be cited as a strong instance in point of the agreement of many eyes and instruments, even by the prevalent rule-of-thumb system; because that object has an advantageous position in altitude when on or near the meridian, to facilitate this branch of inquiry. It will be borne in mind that β is an optical pair; for, as I have elsewhere remarked, strong colours in the *pulcherrima* of the heavens are not at all indicative of motion. To moderate the ordinary complaints about bias, magnifying power, aperture, achromatism, spherical correction, and local atmosphere—and to test the merit of gazing "with all its imperfections on its head"—the eyes and telescopes of various observers, at various places, have been summoned in aid; and the results of the requisition are to the following effect:—

OBSERVER.	DATE.	RECORDED COLOURS.	PLACE.
Sir William Herschel .	1779	A. *Red** B. *Blue*	Slough
———	1781	A. *Pale red* B. *Beautiful blue*	Slough
———	1783	A. *Red, or orange* B. *Blue, or purple*	Slough
Sir John Herschel . .	1816	A. *White, or yellowish white* B. *Blue*	Slough
———	1830	A. *Yellow* B. *Blue (superb)*	Slough
Herschel and South. .	1822	A. *Yellow* B. *Blue (strongly contrasted)*	London
Admiral Smyth . . .	1830	A. *Bright yellow* B. *Fine blue*	Bedford
———	1837	A. *Topaz yellow* B. *Sapphire blue*	Bedford
———	1849	A. *Golden yellow* B. *Smalt blue*	Hartwell
———	1857	A. *Orpiment yellow* B. *Turquoise tint*	Hartwell
Mrs. Smyth	1830	A. *Fine yellow* B. *Blue*	Bedford
———	1857	A. *Orange yellow* B. *Greenish blue*	Hartwell

* For the prevalence of red in Sir William Herschel's star-tints, see the remarks, *ante*, p. 15.

OBSERVER.	DATE.	RECORDED COLOURS.	PLACE.
Struve the elder . .	1832	A. *Yellow (colores sunt insignes)* B. *Cerulean blue*	Dorpat
Benedict Sestini . .	1844	A. *Orange gold* B. *Azure*	Rome
Piazzi Smyth . . .	1856	A. *Pale yellow* B. *Blue*	Teneriffe
———	1862	A. *Yellow* B. *Bluish green*	Elchies
Padre Secchi . . .	1857	A. *Reddish yellow* B. *Green (fine colours)*	Rome
Rev. W. R. Dawes .	1857	A. *Crocus yellow* B. *Greenish blue*	Haddenham
———	1864	A. *Golden yellow* B. *Greenish blue*	Haddenham
Isaac Fletcher . . .	1850	A. *Yellow* B. *Blue*	Tarn-bank
———	1857	A. *Rich yellow* B. *Brilliant blue*	Tarn-bank
———	1864	A. *Rich yellow* B. *Greenish blue*	Tarn-bank
Mrs. Fletcher . . .	1864	A. *Fine yellow* B. *Bluish green*	Tarn-bank
Lord Wrottesley . .	1857	A. *Golden yellow* B. *Greenish blue*	Wrottesley
———	1864	A. *Reddish orange* B. *Clear blue*	Wrottesley
Captain Higgens . .	1864	A. *Golden yellow* B. *Bluish grey*	Bedford
Mrs. Higgens . . .	1864	A. *Golden yellow* B. *Faded sapphire blue*	Bedford
Rev. T. W. Webb .	1849	A. *Fine yellow* B. *Ultramarine blue*	Hereford
———	1864	A. *Full yellow* B. *Aqua cœlestis blue*	Hereford
George Knott . . .	1864	A. *Golden yellow* B. *Bluish green*	Cuckfield
Mrs. Knott	1864	A. *Variable, with ruddy cast* B. *Emerald green*	Cuckfield
Dr. Lee	1832	A. *Orange tinge* B. *Bright blue*	Bedford

OBSERVER.	DATE.	RECORDED COLOURS.	PLACE.
Dr. Lee	1864	A. *Pinkish yellow* B. *Cerulean blue*	Hartwell
Mrs. Lee	1864	A. *Orange* B. *Green*	Hartwell
Sir Rowland Hill .	1864	A. *Yellowish orange* B. *Bluish green*	Hartwell
Lady Hill	1864	A. *Yellow orange* B. *Green*	Hartwell
W. R. Birt	1864	A. *Yellow dashed with orange* B. *Clear pure blue*	Hartwell
Samuel Horton . .	1864	A. *Fiery yellow* B. *Sky-blue*	Hartwell
William Huggins . .	1864	A. *Yellow* B. *Blue*	Tulse Hill
J. R. Hind . . .	1864	A. *Pretty deep yellow* B. *Sapphire blue*	Twickenham
C. G. Talmage . .	1864	A. *Straw-coloured* B. *Blue*	Twickenham
Frederick Bird . .	1864	A. *Pale yellow* B. *Light emerald green*	Birmingham
Rev. Robert Main .	1861	A. *Yellow* B. *Blue*	Oxford
———————	1864	A. *Orange* B. *Sea-green*	Oxford
Mrs. Main	1864	A. *Orange* B. *Sea-green*	Oxford
Rev. G. Fisher . .	1864	A. *Bright orange* B. *Violet*	Hartwell

Replies to my request.

In this enumeration all those observations given after the year 1856 were specially noted at my request—a request always acquiesced in with kind readiness by the parties addressed; and, as some of the replies contain matter in illustration of the recorded facts, the reader shall be treated to a few extracts from them in the order in which they were received.

Sir John Herschel.

Sir John Herschel, in answer to my inquiries, states—under date July 4th, 1864—"From general recollection of the object when looked at, I should certainly say the observation of November 3rd, 1816, does not make the large star enough coloured." And, to a query regarding one of Sir William's epochs, his son has since informed me, that he copied verbatim what he found written as to β Cygni in the registered sheets taken from his father's diaries. The excerpt, which is valuable in this relation, stands thus:—

Sep. 12, '79. Two fine stars. One *red*, the other *blue*.
Sep. 6, '81. The large star (pr.) *pale red*, nisi *pretty red*. The small, a *beautiful blue*. 227; under 460, estimation of colours the same.
Feb. 8, '83. Large, *red* or *orange*. Small, *blue* or *purple*.

The Rev. W. R. Dawes, in a letter of July 5th, 1864, says— "I succeeded last night in getting some very satisfactory observations of β Cygni for the colours of the components. I examined them with powers from 93 to 890. With 93, 153, and 258, I pronounce the colours to be

The Rev. W. R. Dawes.

A. Golden yellow. B. Greenish blue.

"With powers 405, 557, and 890, they appeared

A. Golden yellow. B. Slightly more of a greenish tinge than with the lower powers.

"With one of Horne and Thornthwaite's so-called 'Aplanatics'—which is certainly as achromatic as possible in all parts of the field—power 302, the colours were precisely the same as with the lower powers above quoted. Each star being alternately concealed by a small field in my solar eye-piece, the colours remained the same, so they were also with a perforated diaphragm over the whole object-glass. For comparison I noted the colours of the components of *a* Herculis—

A. Light orange. B. Blue-green.

Compared with β Cygni, A is rather redder, and B rather greener.

"The night was favourable for such observations, being free from haze, and calm. The scarcely varied colours with the highest powers compared with the lower, arises I doubt not from the fine correction of this object-glass, which is not *over-corrected*. In the violently over-corrected Munich ones there is great difference between the high and low powers. Mine is, I think, almost exactly like your old Bedford favourite in respect of correction."

Captain Higgens.

The same post also brought me dispatches of the same date and on the same subject from my friends Captain Higgens and the Rev. T. W. Webb. The Captain informs me, that, "after the spell of cloudy weather which we have had so long, I almost despaired of being able to see the stars last night; but a clear came on before sunset, and as the darkness increased the definition became very good. I think I never saw Jupiter so well before at so low an altitude as I did soon after sunset. As the night advanced objects became at times a little blotty, but on the whole we had a good time of it.

"We had a long scrutiny of β Cygni. My wife sees the pair as follows: A golden yellow; B *faded* sapphire blue; while I make A golden yellow, and B bluish grey. I have no doubt we mean the same tints.

"I am sorry that I have not looked at this object oftener. I see a note in my log of May 24th: A golden yellow, B faintish blue, and I cannot help thinking that the colour of the latter is not so vivid as I saw it last year. I shall keep it well in sight during the remainder of its appearance. We looked at it with 115 to 200 (Cooke's 4-inch, 5-foot focus), which are the powers best adapted for my telescope for ordinary objects when testing colours."

The same night, July 4th, Captain Higgens turned his instrument upon 95 Herculis charged with magnifying powers of 115 and 200; the definition was good, and the normal colours plain, but faintish.

The Rev. T. W. Webb.

The Rev. T. W. Webb entered warmly into the question, saying; "We have had such a prevalence of cloud of late, that, although I determined to take the very first opportunity of attacking your object, I feared it might be long deferred. Last night, however, proved unexpectedly clear, and I set to work at once, and with various powers—65 and 111 Ramsden, 170 Huyghens, 264 single lens, 323 Steinheil, and 600 Huyghens, and "macro-micro"—I had the same result, which was quite independent of your own estimates, as I had forgotten their precise terms. A appeared to me a full yellow, varying to topaz, B an 'aqua cœlestis' blue, with a very slight tinge of green.

"By 'aqua cœlestis'—an old alchymical name which I picked up I know not where—I mean the splendid blue produced by adding ammonia to a solution of sulphate of copper,—ammoniuret of copper is, I fancy, its chemical name. The star reminded me at once of this, only in comparison its blue appeared a trifle greener. This comparison is made from memory only, as I have not seen any of the liquid so called for years, and have not the materials for making it at hand."

Chemical tests are of material use in watching the colours of stars, since they can be made and adapted at all times, and, generally speaking, in all places.

G. Knott.

Mr. George Knott, who complied with my request (July 6th) as "a most pleasant duty," reported in these terms: "On Monday evening at about 11 o'clock the clouds cleared off, and, accompanied by my sole assistant, Mrs. Knott ('an astronomer's best assistant,' as I think the talented Astronomer-Royal for Scotland says somewhere), I adjourned to the observatory with a view of attacking the redoubtable β Cygni with my Alvan Clark refractor. The images of the stars were, however, rather confused, and the state of the atmosphere not favourable to colour observations, for which a *really fine* night seems almost as indispensable as for delicate measurements; but the following are our entirely independent estimates—*valeant quantum*:—

β CYGNI.

7½-inch Equatoreal, full aperture. Powers 173 and 328—Thornthwaite's aplanatic :—

Mr. Knott { A. Full golden, with orange cast.
B. Fine pale bluish green. Decidedly green.

Mrs. Knott { A. Variable, with ruddy cast?
B. Emerald green.

With the aperture contracted to 3·7 inches, and negative eye-piece of power 258, I thought A full golden, with ruddy cast, and B pale blue green; and my estimates were confirmed by Mrs. Knott.

"Last evening was finer, and I had some very good views of this glorious pair with various powers and apertures, and estimated

A. Full clear golden yellow.
B. Delicate pale bluish green.

"With 3·2 inches aperture I thought the colour of B more decidedly bluish, but still with a green cast. In the finder (2 inches in aperture and magnifying power 20) B appeared more decidedly bluish; but in the large telescope, with apertures from 4 to 7·33 inches, its tint was a very delicate and decided *bluish green.*

"To enable you to judge of my colour estimates better, I then examined *a* and 95 Herculis, with the following results:—

a Herculis . A. Golden.
B. Greenish blue, deeper in colour than B of *β* Cygni.

95 Herculis . A. Pale greenish yellow.
B. Pale rosy yellow."

On the 10th of August, Mr. Knott wrote to me again, announcing a fresh testimony —

"Thinking you might perhaps be interested to learn what could be made of the colours of *β* Cygni with one of the *Silver on Glass* reflectors, I wrote to an ingenious correspondent (Mr. Bird of Birmingham) who has a large one of his own make, 11⅝ inches aperture, and of excellent quality (he discovered the duplicity of the small star following Procyon at a distance near the parallel, consisting of two 9-10 mag stars, distant from each other about 0″·6), and asked him to oblige me by examining that glorious pair.

"I received his reply a day or two ago, and now proceed to quote the *ipsissima verba* of his report:—

"I had several looks at *β* Cygni for the colours of the stars, and my impression is, that the large star is a pale yellow and the small one a light emerald green stained with blue. The green is very obvious with a large aperture, and the field seems to be tinted with it. Mrs. Bird does not see the green so plainly, but a young lady staying with us pronounced decidedly for the green, toned with blue. When the stars are put out of focus, the blue seems to be the most prominent. I should esteem it an honour to know that Admiral Smyth was in any way interested in the showing of colours by my silvered-glass speculum and Munich prism, &c. &c."

"Thus the blue green colour of B is quite confirmed. The testimony of a *reflector* is very interesting on the point."

Among those whom I petitioned on this occasion was the Earl of Rosse, for I was desirous of obtaining such evidence also as would be afforded by his noble gigantic instruments. Unfortunately for my object, his Lordship was in England, and a letter from Brighton of the 7th of July thus informed me: "Although I have so often seen β Cygni with both our telescopes, I have never made any memorandum which would enable me to answer your question. In fact I have merely regarded it as a splendid object for visitors who were not astronomers.

Earl of Rosse.

"My assistant, I am sorry to say, is now absent at the sea-side, being very unwell, and I could not rely upon any of the men about the observatory, as they have merely been trained to work the instruments. The first opportunity we will do our best to answer your question accurately."

I have merely inserted this in order to acquaint the reader that, if Lord Rosse's promised appreciation of the colours of this double star comes to hand before the *finis* of this *brochure*, it shall be duly appended.

Mr. Isaac Fletcher (July 9th) writes from Tarn Bank: "Since I received your request to examine the colours of the components of β Cygni the nights have all been cloudy except last night, when at 10 o'clock I levelled the 12-foot refractor at β, and examined the colours with various powers from 134 to 425. I make the colours as follows:—

Isaac Fletcher.

A. Rich yellow. B. Greenish;

and Mrs. Fletcher's independent judgment is

A. Fine yellow. B. Bluish green.

"Afterwards on referring to the 'Speculum,' Mrs. Fletcher said that Mr. Dawes's description of A as 'crocus yellow' was to her mind exactly specified."

Dr. Lee, in a report from Hartwell dated 20th of July,

Dr. Lee.

remarks: "The colours of β Cygni were observed with a power of 118 on the equatoreal instrument, after the stars had been examined under an eye-piece magnifying 50 times. A diaphragm with a wide slit was inserted, by which each star could be examined separately; but no alteration was effected in the estimation of the components as otherwise scrutinized. As you seemed desirous of obtaining the appreciation of various eyes, in addition to those sent to you already, I may add that Miss Wilkinson, and Messrs. Drummond Davis, Douglas Brown, and J. E. Hall, were agreed in pronouncing the colours to be yellow and blue. Mr. D. A. Freeman, however, thinks that the larger star appeared to him red, not deep red; the smaller, under various powers, was light blue.

"With an eye-piece magnifying 240 times, Mr. Birt was impressed with the *great purity* of the blue of the star B, and on calling it to mind afterwards was disposed to regard it as fine *azure*. Mr. Horton compared it to the blue colour of the egg of the hedge-sparrow, certainly a rural,—but very applicable, comparison.

"That the telescope was in excellent working order was evinced by its sharp definition of the clusters in Perseus, the annular nebula in Lyra, and various parts of the moon's surface,—she then being a little after her first quarter. Your two variable stars following 80 Messier have re-appeared, and are duly noted for use."

Sir R. Hill.

Sir Rowland Hill, whom I have long appreciated as an amateur astronomer,* thus described the hues in question: "On the 11th of July, we examined the fine double-star β Cygni, the recorded estimates of colour, with power 118, being as under—

The large star A. Yellowish orange.
The small star B. Green.

"A lower power (50) gave a rather bluish tint to the smaller star; but the concealment of one of the stars by a

* Sir Rowland's scientific bias was developed long before his great and effective POSTAL REFORM; he having been elected into the Astronomical Society in 1822.

diaphragm of metal did not alter the previous impressions; thus proving that the colours really emanated from the stars, and were not merely complementary tints.

Lord Wrottesley.

Besides Lord Wrottesley's colour-notation of July 19th, above recorded, he was good enough to send me the micrometric measures of the angle of position and the distance between the components of β Cygni, as expressly noted by Mr. Hough, his lordship's astronomical-assistant; and which I received on the 21st of that month. Now as these determinations directly confirm the declared opticity of this object, and therefore bear out my dictum that vivid colours are by no means indicative of motion, I will enrol them here:

Mag. and Colour.		Position.	W.	No. of Obs.	Dist.	w.	No. of Obs.	Power.	Date. 1864.
A.	B.								
3 *Pale gold* .	4-5 *Bright blue*	56° 26′	6·9	10	34″·532	6·1	10	450	July 6
Light orange .	*Greenish blue*	NO MEASURES TAKEN.						85	July 8
3 *Light orange* .	4-5 *Light blue* .	54° 28′	6·8	10	34″·405	6·8	10	450	July 13
3 *Golden orange*	4-5 *Light blue* .	55° 39′	5·5	8	34″·255	5·1	8	450	July 14

In the above table W. and *w.* are the sums of the weights assigned to each single observation, divided by 10.

Fixity of β Cygni.

The fixity of β Cygni being rather a standard point to establish in stellar Chromatics, we will repeat the results of the former operations, which are extended by reducing right ascensions and declinations into angles and distances, so that Bradley and Piazzi may be received in evidence. The whole presents a singular view of the agreement of various observers and various instruments in different places, and at different epochs, insomuch that ALBIREO must be deemed one of the best determined stars in the heavens, and therefore very fitting for the object before us. Its place in latitude and longitude is, of course, beyond cavil, it being settled by the meridian labours of many public observatories; while the extra-meridional operations exhibit the following deductions:—

OBSERVER.	POSITION.	DISTANCE	EPOCH.
Bradley . . .	57° 34′	34″ 20	1755·00
Herschel I. . . .	54° 52′	34″·83	1782·45
Piazzi . . .	54° 31′	34″·28	1800·00
Struve I. . . .	54° 30′	34″·29	1821·76
Herschel II. and South	54° 45′	34″·38	1822·98
Dawes . . .	55° 32′	34″·51	1830·54
Smyth, W. H. . .	55° 24′	34″·2	1830·81
——— ———	55° 36′	34″·4	1837·58
——— ———	56° 12′	34″·1	1854·67
Fletcher . . .	56° 25′	34″·412	1850·83
——— ———	55° 41′	34″·370	1857·42
Lord Wrottesley . .	55° 26′	34″·557	1857·47
Padre Secchi . .	54° 57′	34″·419	1858·29
Smyth, C. Piazzi .	55° 28′	34″·50	1862·72
Hough . . .	55° 21′	34″·397	1864·53
Main	56° 00′	34″·28	1864·63

Mean of the above.

The arithmetical mean of which, at the mid-epoch, or 1809·5, gives for the resulting data:

Position = 55° 31″ Distance = 34″·383.

And now I quote myself (*Cycle*, vol. ii. p. 450) in remarking that these conclusions display a very remarkable constancy both in angle and distance, especially as the two components appear to be affected with proper motions, the amount of which does not differ so much in the several reductions, as in the course or direction of the march.* The following are the

* Twenty years after the above was written, my son, discussing β Cygni, says—"In *magnitude*, B is supposed to be slightly variable. In *colour*, there would seem to be yellow pulsations; at their maximum, converting the simple yellow of A into golden yellow, and the blue of B into greenish blue.

"In *position* and *distance*, these stars present one of the most remarkable instances known of near conformity amongst all ages, countries, and kinds of observers. In position this is partly due to the large distance, but not altogether; and in distance, the same largeness should tell rather adversely; so that there is in fact only this explanatory supposition left, if it does explain anything, viz. that, as Professor de Morgan has shown, in his treatise on the Theory of Probabilities, that there must be occasionally amongst men "lucky individuals;" so amongst the double stars, there may be found here and there a happy pair, where no one can well avoid making a good observation."

inferred present values and signs, and the next rigorous comparison may decide between them:

PIAZZI.			BAILY.		
Star A in Æ	— 0s·07	Dec. + 0″·05	Star A in Æ	+ 0s·03	Dec. + 0″·02.
Star B	— 0s·13	+ 0″·04	Star B	+ 0s·05	+ 0″·04.

Mr. Main's recent investigation gives no change in declination for A, and only 0s·002 in Æ: wherefore, considering all the infirmities still adhering to even the best observations, it seems necessary to wait a little longer for absolute conclusions.

Mr. William Huggins, of Upper Tulse Hill, wrote to me on the 30th of July, announcing—" The earliest possible opportunity to attack β Cygni occurred last night. In a small box, which you will receive by the same post as this note, you will find the colours of β Cygni, as they appeared to me, with my telescope, last night. The coloured solutions are to be observed in the dark, by means of the light of a parafine lamp.

W. Huggins.

" The blue pleases me much—with the yellow I am less satisfied. I tried many other yellow-tinted solutions, but with less success. The brighter star appears less yellow when its tint is not heightened in the eye by the simultaneous view of its blue companion. I estimated the colours by viewing each star separately, and should name the tints:—

A. Very dilute bichromate potassa yellow.
B. Dilute ammonio-sulphate of copper blue.

" With my new spectrum apparatus I saw separately the spectra of A and B. They differ, as was to be expected, in as marked a degree as do the colours of the stars to the eye. I refrain from describing the spectra at present, as I wish to re-observe them on an evening when the state of the atmosphere is more favourable. The spectrum of B is exceedingly teazing to the eye. It being of the smallest magnitude which admits of analysis by the spectrum with my telescope of eight inches aperture."

Now here is a proof of what has been advanced respecting

the difficulty of making these most delicate observations, for I can aver that the night of the 29th was what a star-meter would call *beautiful* for micrometric measures, yet the spectrum appliance required the welkin not only to be fine, but superlatively fine. The results, however, when obtained are wonderful: Sirius—a brilliant object for the purpose—displays a spectrum containing five strong bands, and numerous finer lines. The occurrence of sodium, magnesium, hydrogen, and probably of iron, has been negatively detected in that star's atmosphere, which is probably more charged with vapours than that of our sun—a strong argument for the *pluralism of Worlds*.

The changes of colour which stars of the first magnitude are alleged to have undergone (see *ante*, page 16) are hinged upon very inadequate bases; but we are preparing to bequeath such matters to posterity in a more tangible and convincing form than has hitherto been known. The effects of light, absorption, and high refrangibilities are now becoming unveiled.

J. R. Hind.

Mr. John Russell Hind, on the 2nd of August, says, in answer to my inquiry: "I looked at β Cygni this morning, the first opportunity I have had since receiving your note. I make the large star a pretty deep yellow, while I cannot describe the colour of the small one to my vision better than by your term, "sapphire blue. Power 70."

Mr. Hind then continues his letter with an interesting bit of intelligence which, though not stringently necessary here, may not prove altogether out of place.

"I shall write you about your remarkable companion of Procyon,* as soon as I receive an answer from Dr. Luther as to the possible error of the elements of Hebe at the time of your observation carried back from the elements of 1847. Luther has Hebe in hand, and I find by careful calculation that she would be within a short distance of the spot where you saw the *orange-coloured* star, but I believe it is difficult to

* See the *Cycle of Celestial Objects*, page 182; and the *Speculum Hartwellianum*, pages 236-238.

bring her near enough in latitude when the longitudes agree. What a sensation you would have created at that epoch (1833) had you recognised this planet!"

This would have been a pleasing vision, but that—as I have elsewhere declared—Mr. Fletcher's observation of 1850 precludes its indulgence. On consulting this accurate astrometer on the subject, he pointedly says: "It is true my observation is a lopsided one, the distance most unfortunately not having been taken; but I have no doubt whatever, and never had any, that the object observed by you in '33 and by me in '50 is a star—a variable one, no doubt.

"Not long after my observation was made my memory was rigorously ransacked to recall the circumstances attending it. The facts were these: after measuring the position angle, I referred to the Cycle, and, finding my angle nearly the same as yours in 1833, I thought, 'Well, this is only an optical double star. It is not worth while noticing it any further.'

"Some one (I forget who) sometime ago suggested that my measure of position referred to a very distant object in the same direction; but I am certain that it did not, for so great difference in distance from *your star* would at once have rivetted my attention."

On the 4th of August Mr. Hind addressed me again, saying: "Since writing to you, Mr. Talmage, assistant in Mr. Bishop's observatory, who is gifted with a remarkably strong sight (though I cannot say how far that may be of service in this case), has examined β Cygni with powers 70, 108, and 200. He calls A straw-coloured and B blue, but not so deep as sapphire blue."

The Rev. R. Main.

The Rev. Robert Main on the 15th of August brought the great Oxford heliometer to bear on the object specially for this record, he having noticed the colours three years before. "My judgment," he remarks, "is formed by the impression on the eye after forming a single image of each component by making the images cover each other as they now do at the zero, where they are perfectly round. The magnitudes are 3 and 7—a good 7. As I thought you would like to have contemporaneous measures of this most interesting star, I proceeded immediately to measure the distance and position, which are as follows: distance by a mean of 10 measures = 34″·28, angle of position = 56° 00′."

The Rev. G. Fisher.

In a "report" to me of his eye-impression of the hues of the components of β, Mr. Fisher, the well-known arctic astronomer, observes, that to make his estimation of colour be in accordance with Admiral Smyth's chromatic scale—which he only saw after he had recorded his observations in Dr. Lee's book—instead of bright orange he would have stated the colour of A *bright yellow* with a tinge of red. "This," he continues, "affords another instance of the necessity of establishing some well-devised standard scale of colours for such cases, that might be generally accepted by astronomers."

"The magnifying powers used were 50, 118, and 240, all of which were in harmony as to the actual tints: these were negative eye-pieces, but, from accident, we did not use the single lenses you recommended us to do. From a very slight display of prismatic effect on the evening of September the 5th—which although fine was damp—we were led to suspect a slight deposition of moisture on the object-glass; yet the dew-cap was on.

"By the enclosed extract which Dr. Lee has sent, you will find that a few nights since his friend Mr. Norman Lockyer pronounced A to be yellow, and B sea-green."

The finding.

On overhauling the foregoing details, and sifting them through a numerical adjustment, it appears that the conclusive finding to be pronounced upon the colours of the star under trial, in general terms, is—according to the tabulated matter—

A. Yellow.[3] B. Blue.[2]

Now as the deliberate sentence thus adjudged represents the medium hue of many observations, extending over a period of 85 years, and the object is in apparition to us so copiously as to be in working sight for months, it constitutes a desirable target for the Tyro to trim his eye and instrument by, whenever he is bent upon a chromatic onset in that quarter. And as such it carries my recommendation.

So much for ALBIREO, alias β CYGNI.

Colour blindness.

In the first volume of my Cycle of Celestial Objects, pages 302 and 303 (*and see ante, pp.* 15, 16), I instanced what is now termed "colour blindness," or the abnormal peculiarity of certain eyes in their being unable to distinguish colours correctly, and yet capable of proper action in every other use of them. Every one knows those violent cases of it where a person cannot scan green from red, and other such egregious contrasts, and would not admit such an individual's observations of colours at all; but it is by no means so generally known as it should be, that a personal equation of greater or less amount exists in every case, even without wishing to push tintism *ad infinitum.* The reason of the faulty colouring of so many artists by vocation is, that they really are not aware of many of the refinements of colour; their eyes not perceiving them, their fingers cannot render them. In one of the most intense examples, however, of this chromatic personal equation, although the person could not distinguish so bright a scarlet as the flower of the pomegranate from the genuine green of its leaf, I have had abundant proof that his eye was able to perceive brightness, independent of colour, as acutely, if not much more so, than the generality of men.* It should, however, be observed, that there is to the most normal vision a sensible presence of the red element in either violet or lilac, and the various hues indiscriminately termed purple.

* Some cotemporary elders may be surprised at my not having here instanced the hackneyed story of Dalton—Atomic Dalton—how he bought pink stockings in mistake for drab-coloured; how he went to the meeting arrayed in them; how he scandalized the quakers and quaquerettes; and how the elders threatened to read him out of the Society for his gross misdemeanour. Now I made no allusion to this for the simple reason that, on inquiry, I could place no reliance on it, albeit one of my informers was a Royal Duke. It seems merely to have been a *ben trovato* of some wag, who was aware of the aberration of vision in my philosophical and amiable friend.

The Australian telescope.

By thus calling for an increase of energy in the eye-estimation of sidereal colours, it will be seen that we are not insensible to the advantages of the instrumental method, despite all the perplexities which it now labours under; and we may hope that these advantages will not be lost sight of. If it be true that the Government is about to send a large reflector to Australia to observe the southern nebulæ, how desirable that it should also forward another to a tropical region for observing the planets, and for making chromatic observations of the stars. The Australian telescope will have more than sufficient work with the nebulæ, and the planets with their faint satellites will be low down in the north there, while we have them low in the south here; but the equatoreal telescope will have them in its zenith; and it may be elevated on some table-land there far higher into the atmosphere than the Australian one can be. This is a very important matter where colours rather than brightness are concerned: for a want of the latter may be corrected merely by using a larger aperture; but a distortion of the former, once introduced, is utterly irremediable under the existing theory.

The Melbourne telescope.

Since the above allusion was printed in the *Speculum Hartwellianum*, the subject of the Southern Telescope has been repeatedly agitated, with a prospect of eventual success, although it has unfortunately hung fire. In the meantime a very unexpected event occurred. Sir Henry Berkly, Governor of the new and energetic colony of Victoria, wrote on the 23rd of July, 1862, to the Duke of Newcastle, the Colonial Secretary of State, that a commencement had been made for the erection of an astronomical observatory at Melbourne, and that a sum of 4,500*l.* had been voted by the local legislature for its completion; and, as it was also resolved to employ a telescope of greater power than any previously used in the southern hemisphere, he was desirous of obtaining the advice of the Royal Society towards effecting this object.

After much correspondence on this praiseworthy proposal, it was resolved that the telescope to be recommended should be made by Mr. Grubb of Dublin, under the superintendence of Lord Rosse, Dr. Robinson of Armagh, Mr. De la Rue, and Mr. Lassell. The estimate of Mr. Grubb for the probable cost of the instrument, two four-foot specula, a polishing-machine, and a one-horse power steam-engine, was 5,000*l.*; but, as several things in the construction were necessarily experimental, it was put at 6,000*l.* as a safety valve.

Lassell's letter.

These consultations were barely concluded when General Sabine, the President of the Royal Society, received a letter from Mr. William Lassell dated 22nd of July, 1863, couched in these terms:—

On the occasion of my temporary visit to England, I have had the opportunity of looking into some of the correspondence respecting the proposed four-foot telescope for Melbourne, and in consequence I should be glad to be allowed to state that I do not intend to continue my observations with the telescope of this size now erected in Malta, and described in this correspondence, beyond the period of twelve, or at most eighteen, months from the present time; and that, if this equatoreal should meet the requirements of the Melbourne Committee, I shall then be glad to place it at their disposal.

Subsequently Mr. Lassell himself explained, answering a query, that he means to place his grand telescope at the disposal of the Melbourne Committee as a GIFT, though under certain conditions which would satisfy him that it would be wisely and usefully employed; and while it remains to be seen whether this nobly liberal overture is duly accepted, all hands ought to remember Sir John Herschel's logical declaration, "THE DAYS THAT PASS BETWEEN THIS AND WHEN THE SURVEY OF THE SOUTHERN HEAVEN BEGINS, ARE EACH AN IRRETRIEVABLE LOSS TO ASTRONOMY!"

L'Envoy.

We thus close our exhortations with a well-sounded and unmistakable charge, from one of the ablest and most zealous astronomers of the age. This awful warning would seem amply sufficient to make the sons of Urania bestir themselves, at all events; and the amateurs of the present day are not only better furnished with efficient telescopes than were their brethren of yore, but they are also stimulated and aided by trustworthy treatises available to all men—not as of erst written in the *noli me tangere* of crabbed quodlibets in a dead language. The incurring that irretrievable loss to science pointed out by Sir John Herschel, becomes the more heinous since, as just said, the means of observing have become so easily obtainable; and, what is more, we have attained a commanding station from which the labours of our worthy predecessors have rendered a further advance comparatively easy. Indeed there is so broad a basis of former work to rely upon, that we know exactly what to aim at now; and every successive stratum will help us to raise a pyramid of knowledge, the apex of which may finally reach the *summum bonum* of the cultivated mind's expectation.

Appendix.

APPENDIX.

No I.

The instance of the Double Star 95 Herculis, and its marked variations in colour completely proved.

Prelude.

As this object constitutes a remarkable case of Sidereal Chromatics, in which Nature seems to have been caught at her work, some account of the decided mutations of colour recently detected in it becomes necessary; and the story may be related in a few words. It should, however, be premised that we are told (and it is highly probable), that light, when first emitted from the photospheres of the sun and stars "should be in all cases identical, the differences of colour depending upon the differences of constitution of the investing atmosphere:" but the variability of the hues in question is still to be explained, since they cannot be satisfactorily accounted for by any prevailing scientific theory. It is truly wonderful; and at the present status of the phenomenon, we can only ejaculate St. Augustine's *rem vidi, causam non vidi.*

Still it behoves us, like humble neophytes, not merely to wonder and worship, for that line of conduct would not advance us in the adoration due to our omniscient and omnipresent Creator. We must endeavour to understand the glorious phenomena so benignantly opened out to us as portions of a vast design. Some features of nature may be more important for our comprehension than others, yet in our present nescience we should patiently regard them all with equal care, and scrutinise them until their apparent mystery is unveiled. When that happy consummation may occur, it is not for us to anticipate; but in the meantime we recommend the path we have advocated, to be trod with zeal commensurate with the object. An evident change of brightness or of colour, in a sidereal system, may indicate a long period; but it may also prove a chain of pulsations, which, if apparently irregular to the inexperienced, may finally prove to be regularly irregular.

At page 35 of this little work, it is mentioned that my son, Charles Piazzi Smyth, made his observations on the colours of the components of 95 Herculis in the year 1856, under impressions which occasioned much surprise to myself and others. At length, in the autumn of 1862, he, being on a visit to James W. Grant, Esq. of Elchies, on the banks of the Spey, took the opportunity of efficiently re-observing—among others—the pair before us, with that gentleman's fine equatoreally-mounted refracting telescope, of no less than 11 inches aperture. The following are the conclusions arrived at by the interesting scrutiny; but to a general reader it may be necessary to explain that, under the heading "Authority," σ means Struve using a small instrument, and Σ the same gentleman with the Dorpat 9-inch; "Cycle" and "Spec. Hart." are from my observations, and "Guajara" and "Elchies" are those of C. P. Smyth:—

95 HERCULIS. Approx. R.A. = $17^h\ 55^m\ 33^s$;
Decl. + 21° 35′ 56″ for Jan. 1, 1860.

Registered colours.

Components.	*Magnitude.*	*Colour.*	*Position.*	*Distance.*	*Date.*	*Authority.*
A B	5 5	—— ——	AB 261° 54′	″ 6·28	1822·69	σ
A B	4·5 4·5	Greenish yellow Reddish yellow . .	AB 262° 30′	6·19	1828·71	Σ
A B	5 5	Greenish Yellowish . . .	AB 262° 12′	6·11	1828·76	Σ
A B	5 5	Yellow green Egregiously red .	AB 261° 12′	6·07	1829·62	Σ
A B	5 5	Yellow Yellow	AB 261° 6′	5·88	1832·53	Σ
A B	5·5 6·0	Light apple green Cherry red . . .	AB 261° 48′(*w* 9)	6·1(*w* 9)	1833·78	Cycle
A B	— —	Gold yellow Gold yellow . . .	—— ——	—— ——	1844·5	Sestini
A B	— —	Pale green Reddish	—— ——	—— ——	1851·3	Spec. Hart.
A B	— —	Greyish white Greyish white . .	—— ——	—— ——	1856·58	Guajara
A B	— —	Light green Pink	AB 261° 11′	6·15	1857·42	Fletcher
A B	— —	Greenish Reddish	AB 260° 11′	6·09	1857·45	Wrottesley
A B	— —	Light apple green Cherry red . . .	AB 260° 30′(*w* 7)	5·6 (*w* 6)	1857·63	Spec. Hart.
A B	5·7 6·0	Yellowish with blue tinge Slightly reddish yellow	AB 261° 4′(2)(*w* 1)	6·29(2)(*w* 0·6)	1862·72	Elchies

"The *magnitudes* may vary through half a magnitude.

Remarks by C. P. Smyth.

"The *colours* vary astonishingly; and, as one of the most remarkable cases in the heavens, may be treated at length. On my return from Teneriffe I had communicated the observed colours of this, amongst a series of other stars, to my father and his friends, and they seized on my equality of the colours of the two components of 95 Herculis, as the one great anomaly in my list, and chief divergence from the Cycle, where they are extremely contrasted; and, several observers being called together the next summer, they proved unanimously that the Cycle was right. I remained, however, positive that I had observed the alleged equality, and on two nights; for, being surprised, and in a manner amazed at it, having promised on the strength of the printed description to show a bystander a case of remarkably contrasted colours, I laboured to try to perceive some approach to them; but, though certain strange flickerings of colour appeared on one of the nights, there was nothing to disturb the general equality of the two discs, and their close approach to white, or a light neutral grey.

Its real changes of hue.

" All this time it is true that Sestini's observation of 1844 was on record, pronouncing both the stars gold-yellow, but it was not until recently referring to Struve's original observations, which are partly given above, that I became convinced of this being a pair of cosmical and brilliant changers of colour; A passing

from yellow
to greyish, from that
to yellowish with blue tinge, from that
to greenish, from that
to light green, from that
to light apple green, from that
to "astonishing yellow green," and from that
to yellow again,

"in a period of probably twelve years, while B in the same time passes

from yellow
to greyish, from that
to yellowish with reddish tinge, from that
to reddish, from that
to cherry red, from that
to " egregious red," and from that
to yellow again.

"Real physical changes, these, in each star, there can be no doubt; and, though in this case the one colour is so generally the complement of the other, a case will be found farther on where one component of a double star goes through brilliant periodic changes of colour, without the other component very sensibly altering its tint.*

Is it a binary system?

"In *position* and *distance* 'fixity' and an optical character have hitherto been both assigned and confirmed to 95 Herculis; but now, with the assistance of the Elchies observation and the B.A.C. proper motion, which does *not* make its appearance relatively between the two components, I have the utmost confidence in declaring the pair to be binary, and undergoing a slow retrograde movement in angle, of about 2° in forty years, roughly—a conclusion which, though it must prove fruitless for ages in enabling the orbit to be approximated to, is nevertheless of vital consequence in studying the strange chromatic problem which the stars present."

Notice from Bedford.

While the above exposition was being prepared for the press, the changes in hue of 95 had been noted entirely independently at Bedford, by Captain John Higgens, who favoured me with the details before he could have possibly heard of the operations at Elchies. His statement is printed in the "Monthly Notices" of the Royal Astronomical Society for November 1863, from which publication it is here extracted:

On the Colours of the two Component Stars of 95 *Herculis.*

Mr. Higgens, in a letter dated Bedford, 17th August, 1863, addressed to Admiral Smyth, writes as follows:

"I beg to forward a few notes on 95 *Herculis.*

"When I last saw the star, in the autumn of last year, the colours were as noticed in the *Cycle:* A, apple-green; B, cherry-red.

"I first saw it this year on the 23rd April, but the bright hues were not there, and greenish-white and pinkish-white were all I could affirm.

"May 10th. Hues more faint. I could only record them as dull white, both A and B.

* This alludes to that most interesting object 70 Ophiuchi, the B of which my son assumes to vary through a variety of colours, at periods yet unascertained; and that yellow, green, violet, purple, and red, are permanently registered. But from many conditions of the case, I am persuaded we must await more observations before conclusions can become at all satisfactory.

"Aug. 1st. A, greenish-white; B, yellowish: both changing nightly till Aug. 12, when they showed as in the *Cycle;* A, apple-green; B, cherry-red. A first shewed signs of deepening colour, the hues becoming more apparent every night, B changing from yellow to red more rapidly.

95 Herculis.

"It was not possible to note night by night the amount of change; but it was very palpable after an interval of three or four nights, and continued so till both stars showed their normal colours.

"The instruments used weie, a $3\frac{1}{4}$-inch achromatic with 80, and a 4-inch with 115; both glasses by Cooke, of York."

Admiral Smyth remarks:

"The star in question, 95 *Herculis*, is now a crucial instance of sidereal colour-changing; and it has given some little trouble, both to my son Piazzi and myself already, in poring over the registered observations.

"It is to be hoped that a field of research at once so elegant, easy, and useful, will be followed up by some of those Fellows of the Society who possess both means and leisure to pursue the interesting inquiry."

"The Astronomer Royal, however, suggested to Admiral Smyth that the simultaneous change for the two stars is suspicious, and looks like a possible change in the telescope."

Epoch of 1864.

Captain Higgens continued his watch over this object, during the apparition of the present year, with commendable perseverance; and on the 10th of August wrote thus:—

"I beg to forward the results of thirty-eight observations of 95 Herculis, made on different days from May 21st to August 8th, with Cooke's achromatic telescope of 4-inches aperture and 5-feet focus, under powers 115 and 200.

"Of the thirty-eight observations eight are noted as simply normal, nineteen as shewing A fainter than last year when in full normal colour, and eleven as shewing both A and B fainter; and my impression is, that they are not so vivid as they were last September, and that A shews more loss of colour than B, though neither to the extent of justifying their being classed as anything but normal, though somewhat faintish.

"It has always struck me that the definition of their colours in the Cycle, a slight apple-green and cherry-red, is the most graphical description that could be applied to these beautiful objects. I have watched them with great care and interest, from having had the good fortune last year of detecting the change of colour from white to their normal condition, as reported to you."

On the 11th the Captain informed me, "I received a letter from the Rev. Mr. Webb this morning, in which he incidentally mentions 95 Herculis as follows:—'I fancied the other night the red decidedly more marked than the green.' Now, as I had not mentioned to him that I have been engaged in observing 95 Herculis *this year*, I feel pleased at so independent a confirmation of my notation of A (*green*) being fainter than usual."

APPENDIX NO. II.

THE COLOURS OF DOUBLE STARS AS AFFECTED BY THE ANOMALIES OF RAYS ARISING FROM THE TREMORS—AND OTHER EFFECTS—OF THE ATMOSPHERE.

Of the atmosphere.

In descending to the lower strata of the atmosphere, the reader should be informed that we here restrict the term to the portion of ambient air next the Earth, which receives vapours and exhalations, and increasingly refracts the rays of light. Ptolemy had already remarked that the light of the stars underwent a change of direction in the course of passing through this terrestrial envelope; and every observer is aware of its darkening effect, especially under high magnifying power. It therefore follows, that the pervading consequence of angular altitude is a paramount condition to be considered before the amateur expends much time upon objects near the horizon, wherever he may happen to be located. On this account the diatribe was uttered against low stars (see *ante*, page 46); though perhaps an occasional dip among the remote regions of the South may hereafter yield comparisons in aid of further investigations of the phenomena—visible and invisible—of vapours.

A conviction of this will very soon come home to the intelligent observer, who will shape a course accordingly, trimming agreeably to his means and intentions; and his judgment must guide him in selecting objects for a systematic attack. Not even will a tropical climate free the spectator from these tremulous deceptions. On the contrary, the absence of aqueous vapour may be perceived by anybody to diminish the apparent steadiness of objects, in proportion to the amount of heat and distance from the eye.

Meantime, as it is the most able discussion of the point that I am aware of, I will here extract my son's generalization of the argument for our climate from the twelfth volume of the Edinburgh Astronomical Observations, page 466.

"When we look at the broad discs of either moon or sun in the act of rising or setting, the colouring effects of the atmosphere are plainly visible on them, as every painter knows right well; and the same effects we might expect to witness on a star seen in a telescope under the same small angle of altitude; but, when we do actually look into such an instrument, the adventitious colour of the region for a surface of sensible size is generally overpowered by the prismatic effect on the mere point presented by a star, which is thereby converted almost entirely into a spectrum of red on one side and blue on the other. Of course the observer attempts to eliminate these prismatic tints from, or rather takes no note of them at all in, the true cosmical colour of the centre of the stellar image; but does he succeed in eliminating both these and every other colouring effect of the atmosphere? To ascertain this point, I have taken all the stars observed on Teneriffe in south declination, and, having strengthened them from some observations by Mr. J. W. Grant at Calcutta, and by Sir T. Maclear at the Cape of Good Hope, have compared them, in the matter of colour, with the best observations of north and middle Europe; and then, dividing them into three groups, of three different degrees of more and more southerly position, we have as below:—

Argument of C. P. Smyth.

Influence of atmosphere at low altitudes.

COLOURS OF STARS AS AFFECTED BY ATMOSPHERE.

Decl. 0° to − 10°.

Star.	*Components.*	*Magnitude.*	*European Colour.*	*Teneriffe and Indian Colour.*
113 P. Ceti . . .	A	7	Yellow	Strong yellow
	B	9	Blue	Warm grey
146 P. Ceti. . .	A	6·5	Yellow	Yellow
	B	9	Violet	Pale violet
32 Eridani . . .	A	5	Orange	Orange
	B	7	Greenish	Greenish
5 Aquilæ . . .	A	7	White	Pale yellow
	B	8	Blue lilac	Bluish
ψ^1 Aquarii . . .	A	4	Orange yellow	Cadmium yellow
	B	4·5	Blue	Blue

COLOURS OF STARS AS AFFECTED BY ATMOSPHERE.

Decl. – 10° to – 20°.

Star.	Components.	Magnitude.	European Colour.	Teneriffe and Indian Colour.
ν Serpentis . . .	A	4·5	Greenish	Bluish
	B	9·0	Coppery	Warm lilac
185 P. Antinoi . .	A	9	Yellow	White
	B	10	Blue	White
	C	12	Violet	Blue
186 P. Antinoi . .	A	7·5	Whitish	Yellow
	B	9	Blue	Blue
α^2 Capricorni . .	A	3	Yellow	Yellow
	B	16	Blue	Blue
τ^1 Aquarii . . .	A	6	White	Light yellow
	B	9·5	Violet	Pale violet
94 Aquarii . . .	A	6	Orange	Yellowish
	B	8·5	Greenish blue	Faint lilac
107 Aquarii . . .	A	6	White	Pale yellow
	B	7·5	Bluish	White

COLOURS OF STARS AS AFFECTED BY ATMOSPHERE.

Decl. – 20° to – 30°

Star.	Components.	Magnitude.	European Colour.	Teneriffe, Indian, and Cape Colonr.
α Scorpii . . .	A	1	Fiery red	Orange red
	B	11	Emerald	Pure blue
39 Ophiuchi . . .	A	5·0	Orange	Pale yellow
	B	7·5	Bluish	Faint blue
α Piscis Austr. . .	A	1	Reddish	White
	B	9·5	Dusky blue	Blue

Atmospheric effects perceptible.

"Now, from the first of these lists, it would appear that all its five pairs of stars are coloured to European, almost precisely as to more southern, observers; and that, in so far, European observations of colour are quite safe down to 10 degrees below the celestial equator.

But in the second list, or from 10 to 20 degrees below the equator, three out of its seven stars show unmistakeably symptoms of atmospheric colouration—yellowish being changed into orange, bluish into greenish, and white into blue. While in the third list, or from 20° to 30° south declination, every star there shows, and in each of its components, the same effect to an increased degree—white being turned into reddish, pale yellow into orange, orange red into fiery red, and pure blue into bluish green, or dusky blue, as depending doubtless on variations of the lower atmosphere at the several times of observation.

Safe limits for colour-observations on stars.

"Hence there can be hardly a doubt but that, in the important physical question of the real colours of stars, no European observations should be received below the fifteenth degree of south declination; and that, if powerful spectrum analysis be employed, the range should not exceed the fifth degree, or just so far beyond the equator as may give a southern observatory, attending to the southern stars, a few common objects of observation for index differences.

"Meanwhile, with regard to the Teneriffe and Elchies stars, it is most satisfactory to find, after throwing out the cases of atmospheric disturbance, that the rest are, on the whole, so very trustworthy and similar to both the 'Cycle' and F. W. Struve's colours, that, if a case of particular divergence be therein found, it is capable of being referred to a physical change of the star's real colour; almost as confidently indeed as an eye observation of change of brightness may be assigned to a real change of 'magnitude;' and, consequently, renders the exact date of the chromatic phenomenon a necessary accompaniment to its description."

Postscript.

On Nebulæ.

ALTHOUGH the cause is utterly unknown, and in the present stage of human cognoscence appears to be inscrutable, it is surmised that the exceptional bodies designated Nebulæ have a connection with double-stars (*see Arago's Popular Astronomy, book xi. chapter xxiv.*) while, as to colours, I have noticed in them pale tints of white, creamy white, yellow, green, and blue. It therefore follows that these incomprehensible but palpable evidences of Omnipotent power and design are not unnecessarily hauled in and appended to our dissertation upon Sidereal Chromatics.

It will be recollected by all who are really concerned about the matter, that, when the wondrous revelations of Lord Rosse were communicated to the public, certain buzzing popinjays, who hang about and obstruct the avenues to the temple of science, vociferously proclaimed that the Nebular Theory had received its *coup de grace* from the castle at Parsonstown. Now this crude conceit was assuredly not imbibed from his Lordship's statement, he having most pointedly said, that "now, as has always been the case, an increase of instrumental power has added to the number of clusters at the expense of the nebulæ properly so called; still it would be very unsafe to conclude that such will always be the case, and thence to draw the obvious inference that all nebulosity is but the glare of stars too remote to be separated by the utmost power of our instruments."

In the *Speculum Hartwellianum* (pages 111-114) I gave my fully considered opinion on this head, showing the actual state of the question, and advocating that planetary nebulæ diffuse patches of we know not what emanating light, and all the "island universes" so profusely scattered in the abysses of space, should be competently watched for ages. Now those "Thoughts on the Nebular Hypothesis" were written in the year 1860, as an addendum to what I had already published on the same subject in my *Cycle of Celestial Objects.* It was therefore with pleasure, while this, the last sheet, is passing through the press, that I received a letter from my unflagging friend Mr. William Huggins, in which he thus announces the conclusion arrived at from the masterly experiments lately made at Tulse Hill: "I fancy you will be interested in the result of some observations I have recently made on the *spectra of some of the nebulæ*. I have obtained evidence, which I believe will be accepted as satisfactory, that certain of the nebulæ (at present my list contains five PLANETARY nebulæ and the annular one in Lyra) are *certainly* NOT *clusters of stars*. They are probably enormous masses of glaiding or ammoniacal gas containing comparatively small quantities of matter condensed into the liquid or solid state. The observations are now at the printer's, and I hope within a fortnight to send you a copy of them. After the opinions which you have published, thinking you must be greatly interested in the matter, I have ventured to trouble you with this note in anticipation."

On Nebulæ.

So enchanting a vista of successive discoveries has of late been thrown open to us, through following up the long-neglected unravelling properties of the Prism, and applying the delicate yet unequivocal test to the heavenly bodies, that a few familiar explanations of the spectra from so experienced a hand as Mr. Huggins may be acceptable to many an amateur-gazer who is now endeavouring on fine nights to interpret those brilliants. In expounding the subject, he thus popularly expresses himself:—

Mr. Huggins's explanation.

"The dark spaces due to interference are supposed to be produced by the action of light upon light—but the dark lines of the spectrum by the absorptive action of vapours or gases upon light. In the case of interference according to the undulatory theory, when two waves of homogeneous light from the same source, and proceeding by two different routes a little unequal in length, meet in opposite phases of an undulation, they destroy each other's motion, and so there results a calm in the luminiferous ether. When no waves dash upon the retina, the eye is without stimulus, and this calm we call darkness.

"The present theory of the nature and origin of the dark lines of the solar spectrum is quite recent. In 1858 Balfour Stewart published a paper on the law of exchanges in radiant *heat*. In the following year the subject was taken up by Kirchhoff, who extended the theory to *light*. The conclusion at which he arrived may be thus stated:—When any substance is heated or is rendered luminous, rays of certain and definite degrees of refrangibility are given out by it, whilst the same substance has also the power of absorbing rays of these identical refrangibilities. Thus the light of burning sodium or glaiding, sodium vapour, cannot get through the vapour of sodium, though this vapour is wholly powerless to absorb or quench light of any refrangibility other than that which sodium vapour emits when heated till it is luminous. In this way the double line D of the solar spectrum can be experimentally produced.

"How is light absorbed by gas or vapour? The theory is as follows:—The

Mr. Huggins's explanation.

atoms of a *gas* being freer to move than those of a liquid or a solid, are capable of swinging or vibrating at certain definite rates only. These atoms have also the power of intercepting the waves which were excited by atoms swinging at the same rates, and these only. Thus atoms, the motions of which are suitable for the emission of red light, will stop red light, and so on.*

"The maps accompanying my paper show the several distinct rates of vibration corresponding to the motions of the gaseous atoms of the elements described. Each line by its place in the spectrum indicates a definite wave-length, just as each note of a piece of music indicates a definite rate of vibration of the air. These maps may be termed the light-songs of the elements written down in notes. Absorption is therefore according to this theory a transference of motion from the ether to the material particles immersed in it.

"Thus it appears when the light emitted from the solid or liquid photosphere of the sun has to traverse its atmosphere, crowded with vapours of different substances, each vapour stops its own group of lines of light, and so the original light reaches us lessened by the aggregate of all the groups of all the vapours. Now, when this light is spread out by the prism, the dark doings of these vapours are revealed, and dark lines or dark spaces show the places where the light has been intercepted."

Chemical elements.

Such are the reasonable expectations of observational and experimental science, as connected with the constitution of those substances which optical analysis afforded the means of discovering in the heavenly bodies. A reading of Mr. Huggins's statement to the Royal Society on the "Spectra of some of the Chemical Elements" will show the care and labour involved in constructing the Spectroscope apparatus with which he scrutinized gold, silver, thallium, cadmium, lead, tin, bismuth, antimony, potassium, arsenic, palladium, lithium, strontium, platinum, tellurium, osmium, rhodium, iridium, manganese, chromium, cobalt, nickel, and iron. In fine: if the pestilent earth-annihilating prophets who have recently frightened some old women, of both sexes, out of their wits, will but permit our globe to roll noiselessly on its axis for another

* When two waves meet in the relative position shown in No. 1, in consequence of the opposite direction of their motion they are both destroyed. But when two

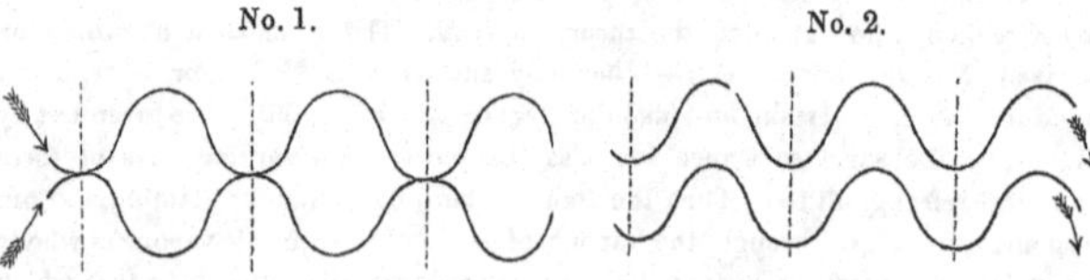

waves meet in the relative position shown in No. 2, they co-operate, and there is a double amount of light.

See also Professor Tyndall on "Heat as a mode of Motion."

century, astronomy will be found in possession of many realities which, in 1864, are only to be classed among the numerous desiderata of Uranology.

MORE ABOUT ALBIREO.

In employing various eyes and instruments upon this fine double-star, I was, of course, desirous of securing the evidence afforded by the powerful achromatic telescope at Greenwich. But it was only as this sheet is being concluded—considerably after the eleventh hour—that I received the reply; yet it is of that interest to the question, that I here append it as received.

β Cygni.

Royal Observatory, Greenwich, London, S.E.
1864, September 29, evening.

My dear Sir,

Various circumstances have impeded the examination of β Cygni, but here I send it at last. You may depend upon the result as accurate; it has been carefully referred to an accurate standard.

I give it, on the other leaf, in the words in which it was given to me.

I am, my dear Sir,
Yours most truly,
G. B. AIRY.

Admiral W. H. Smyth.

Examination of the Colours of β Cygni,
September 28, 1864.

Observations by Mr. J. Carpenter.

"The large star is bright yellow, about the colour of that part of the solar spectrum situate at a point about one-eighth of the distance between Frauenhofer's D and E, reckoning from D towards E. (Sensibly the same colour as the flame of a hand-lamp fed with Colza oil).

"The small star is pale blue, about the colour of that part of the β Cygni.

spectrum which is crossed by line F. (Sometimes it was suspected that the light of this star had a slightly greenish tinge.)

"These comparisons were made on two evenings with a drawing of the spectrum by M. Chevreul, in his 'Exposé d'un Moyen de définir et de nommer les Couleurs,' and this drawing was afterwards compared with the solar spectrum itself at the points used for the star comparisons, and was found to be accurate.

"When the eye-piece was pushed *within* the proper focus, the contrast between the colours of the discs seemed more striking than when the proper focus was used; and, when the eye-piece was pulled out *beyond* the proper focus, the contrast seemed less striking.

"The colours were most strongly contrasted with low powers, as they were also when the image was viewed without any eye-piece. With higher powers the colours became a little more nearly similar, the yellow star seeming to retain its colour, but the blue becoming a little yellowish in its centre."

The ending.

One word in conclusion. With all my admiration of the marvellous and extensive power of Chemistry in disintegrating the nature and properties of the elements of matter, I really trust it will not be exerted among the Celestials to the disservice or detriment of measuring agency; and this I hope for the absolute maintenance of GEOMETRY, DYNAMICS, and pure ASTRONOMY.

INDEX.

Printed in the United States
By Bookmasters